Lecture Notes in Computer Science 15457

The series Lecture Notes in Computer Science (LNCS), including its subseries Lecture Notes in Artificial Intelligence (LNAI) and Lecture Notes in Bioinformatics (LNBI), has established itself as a medium for the publication of new developments in computer science and information technology research, teaching, and education.

LNCS enjoys close cooperation with the computer science R & D community, the series counts many renowned academics among its volume editors and paper authors, and collaborates with prestigious societies. Its mission is to serve this international community by providing an invaluable service, mainly focused on the publication of conference and workshop proceedings and postproceedings. LNCS commenced publication in 1973.

Paul Robertson · Olivier Georgeon
Editors

Situated Self-guided Learning

4th International Workshop, IWSSL 2024
Oxford, UK, September 12–13, 2024
Proceedings

Springer

Editors
Paul Robertson
Dynamic Object Language Labs, Inc.
Lexington, MA, USA

Olivier Georgeon
Catholic University of Lyon
Lyon, France

ISSN 0302-9743 ISSN 1611-3349 (electronic)
Lecture Notes in Computer Science
ISBN 978-3-031-96324-7 ISBN 978-3-031-96325-4 (eBook)
https://doi.org/10.1007/978-3-031-96325-4

This Springer imprint is published by the registered company Springer Nature Switzerland AG
The registered company address is: Gewerbestrasse 11, 6330 Cham, Switzerland

Preface

In this fourth International Workshop on Situated Self-guided Learning (IWSSL 2024), we continued to explore an expanding field of research that attempts to build systems that can learn without human intervention and with little or no hard-wired domain knowledge, as would a newborn child or animal. The theme for this workshop was learning groundings. We explored how to ground learned objects efficiently in online self-guided exploratory learning.

Here are the key points from the workshop:

- **Workshop Theme:** Learning groundings in online self-guided exploratory learning.
- **Research Goal:** Building systems that can learn without human intervention and with little or no hard-wired domain knowledge.
- **Focus of Research:** Situated Self-guided Learning (SSL) for intelligent systems like robots and cyber-physical systems.
- **Workshop Goal:** To discuss the combination of neural and symbolic methods for practical AI applications.
- **Workshop Participants:** A mix of academic and industrial researchers with expertise in both symbolic and neural techniques.
- **Key Issues Addressed:** Grounding, neuro-symbolic approaches, state-of-the-art assessments, and broader AI measurement.

The workshop was invitation only, which guaranteed high caliber attendees. The review process had three reviews per paper, single blind. Nine papers were promised, 6 papers were submitted of which 6 were accepted, one being the introduction to the workshop.

September 2024

Paul Robertson
Olivier Georgeon

Organization

General Chair

Paul Robertson	Dynamic Object Language Labs Inc., USA

Program Committee Chairs

Paul Robertson	Dynamic Object Language Labs Inc., USA
Olivier Georgeon	Catholic University of Lyon, France

Steering Committee

Milan Minsky	Leela AI, USA
Paul Robertson	Dynamic Object Language Labs Inc., USA
Olivier Georgeon	Catholic University of Lyon, France
Kristinn Thórisson	University of Reykjavik, Iceland
Robert Laddaga	Dynamic Object Language Labs Inc., USA

Program Committee

Paul Robertson	Dynamic Object Language Labs Inc., USA
Olivier Georgeon	Catholic University of Lyon, France
Kristinn Thórisson	University of Reykjavik, Iceland
Robert Laddaga	Dynamic Object Language Labs Inc., USA
Henry Minsky	Leela AI, USA
Steve Kommrusch	Leela AI, USA

Contents

IWSSL2024: Introduction to this Volume 1
Paul Robertson and Olivier L. Georgeon

Concept-Centered Knowledge Representation: A 'Middle-Out' Approach Fusing the Symbolic-Subsymbolic Divide 10
Pei Wang and Kristinn R. Thórisson

Society of LLMs: Constructivist Learning Approaches with Multiple LLM Instances 53
Steve Kommrusch and Henry Minsky

Schema Mechanisms 2.0 for Developmental Artificial Intelligence 62
Olivier L. Georgeon, Filipo S. Perotto, Kristinn R. Thórisson, Arash Sheikhlar, and Paul Robertson

Intrinsic Motivation for Artificial Agents 88
Paul Robertson and Olivier L. Georgeon

Intelligence Tests for AI 121
Robert Laddaga and Paul Robertson

Author Index 151

IWSSL2024: Introduction to this Volume

Paul Robertson[1(✉)] and Olivier L. Georgeon[2]

[1] Dynamic Object Language Labs, Lexington, MA 02421, USA
paulr@dollabs.com
[2] UR CONFLUENCE: Sciences et Humanites (EA 1598), UCLy, Lyon, France
ogeorgeon@univ-catholyon.fr

Abstract. This introduction serves to describe the goals of the workshop and put the resulting papers that form this collection into the context of the state of machine learning as it existed in September 2024.

1 Introduction

On 11 September 2024, a small group of A.I. scientists arrived at St. Stephens House in Oxford, England, for a reception on the eve of the fourth international workshop on situated self-guided learning. Self-guided refers to the manner in which learning occurs without direct supervision by exploration guided by intrinsic motivation and its own predictions or heuristics to improve its understanding over time. Situated refers to a system that is embedded within a dynamic environment where it interacts with real-world conditions.

In contrast to previous years, the 2024 meeting occurred at a pivotal juncture. The sustained progress in Large Language Model (LLM) advancements was gradually waning, and efforts to effectively deploy LLMs encountered mounting challenges. The notion that Artificial General Intelligence (AGI) would be imminent appeared increasingly improbable, underscoring the urgent need for novel breakthroughs. While Good Old-fashioned Artificial Intelligence (GOFAI) faced criticism for its fragile nature, artificial neural networks (ANNs) demonstrated unreliability. Object recognition failures often manifested in severe consequences. The propensity of LLM for hallucinations made them untrustworthy. Furthermore, numerous other shortcomings impede the practical application of these systems in diverse domains. A notable example is catastrophic forgetting, where incrementally adding new information leads to the loss of a substantial amount of previously acquired knowledge.

All of the aforementioned achievements represent remarkable achievements for Artificial Intelligence (AI) Large Language Models (LLMs) and Artificial General Intelligence (AGI). Coinciding frequently, these advancements often make unsubstantiated claims that progress in the former will soon lead to the realization of the latter. Although we acknowledge the value and potential necessity of AGI, we firmly believe that the current trajectory will not lead to its achievement, either in the short term or in the long term, without substantial breakthroughs. The papers presented in this volume, which originate from our

P. Robertson and O. Georgeon (Eds.): IWSSL 2024, LNCS 15457, pp. 1–9, 2025.
https://doi.org/10.1007/978-3-031-96325-4_1

Fig. 1. Auguste Rodin's statue "The Thinker" This image was generated by DALL-E using the prompt "Make an image of a robot assuming the same pose as the famous statue of "The Thinker" by Auguste Rodin."

workshop, attempt to explore strategies for accelerating progress toward the achievement of AGI. Learning has been the driving force behind most of the recent remarkable achievements, and we believe that learning will be crucial to achieving AGI. In particular, we emphasize the essential role of developmental learning and the importance of embodied online learning.

What will success entail? It may resemble Rodin's bronze statue of the thinker, but with a robot that replaces the human depicted in Fig. 1. The original statue portrays a nude man whose right hand supports the chin in a posture that conveys deep thought and contemplation. These characteristics, among others, are what we have come to recognize as human intelligence. One day, our AI-powered robots may find themselves in a similar pose.

The press often employs hyperbole to suggest that the advent of Artificial General Intelligence (AGI) is imminent. However, most of the participants in this workshop, if not all, have a different perspective. Although AGI is not a precisely defined term, it is generally understood to encompass intelligence comparable to human intelligence in all dimensions and not limited to restricted domains.

A notable example of non-general intelligence is Alpha-Go [5], which achieved remarkable success through self-play. However, it is essential to recognize that Alpha-Go's capabilities are predetermined by human programmers.

Human and animal intelligence has the remarkable ability to address challenges that were not yet present during brain evolution. We aspire for AGI to possess the same level of adaptability and potential to surpass human capabilities in various domains. To facilitate this effort, it would be beneficial to establish a universally accepted definition of AGI. Although several definitions have been proposed, none have yet gained widespread acceptance. The purpose of this workshop was not to define the term AGI, nor is it the focus of this volume. Instead, we have designed this volume to explore innovative approaches to research and broaden the research agenda. We anticipate that this volume will be valuable to people new to developmental learning, which is why we have included comprehensive literature surveys to facilitate their access to relevant information.

The learning process in animals and humans is a product of a long evolutionary history, resulting in outcomes that seamlessly align with their respective environments. Although humans perceive this learning as intelligent, contemporary LLMs lack the same level of critical comprehension when scrutinized. The primary objective of developmental learning is to achieve the learning capabilities of humans and animals. LLMs have achieved remarkable progress in a relatively short period, but there is a risk that their development veers off course. Text generated by LLMs conveys the impression of understanding, but meticulous experimentation reveals the absence of genuine comprehension. Instead, LLMs possess the ability to present information learned from training data without underlying understanding. LLMs can generate responses that appear to be correct, but this is merely a matter of assembling fragments of information that have previously been encountered. Similarly, generated images exhibit striking similarities to images from training data, and this phenomenon extends to generated code. Many questions can be answered without necessitating comprehension, leading to misinterpretation of understanding and resulting in exaggerated claims about the capabilities of LLMs.

Although these capabilities remain valuable, their ability to generate convincing responses, text, code, or images that are erroneous places a substantial responsibility on the user for verifying the output. Reliability is not the only issue; these systems lack the ability to self-improve. For numerous applications, the ability to acquire knowledge through trial and error is a fundamental aspect of intelligence. Consequently, substantial financial investments are being made to explore safeguards for the responsible utilization of these technologies. In all probability, the inherent risks will persist.

An increasingly large number of individuals are already using these technologies, and this trend is likely to continue. This compilation of papers does not intend to diminish the accomplishments of these approaches. It is highly probable that we have all employed them in limited capacities to generate code, rewrite text, and even generate images. Why not? As long as we take the necessary time to ensure the usability of the result, if it is frequently usable, it can undoubtedly simplify our lives. Throughout history, science has consistently produced useful technologies. Microelectronics, a product of quantum mechanics, and an ever-expanding array of medical treatments derived from genetics are just a few examples. As we progress towards the realization of true and complete artificial intelligence, it is entirely reasonable and anticipated that there will be technologically beneficial spin-offs.

Early artificial intelligence (AI) was primarily focused on reasoning and the targeted aspects of intelligence perceived as the most human, such as our capacity for rational, logical, and complex action planning. In contrast, perception appeared to be less reliant on reasoning and more focused on pattern recognition. ObjectNet and the 2012 competition [3] made significant and abrupt advancements in object recognition, despite limitations. Despite these advancements, over a decade of intense research, object recognition has been widely adopted in numerous applications. Surprisingly, the challenges encountered in early convolutional neural networks (CNNs) continue to exist to this day, partly due to the tendency of many people to adopt the prevailing state-of-the-art without adequately seeking improvements that improve the reliability of the technology.

Since the advent of artificial intelligence, computer vision and natural language processing have been recognized as crucial components of intelligence. For several years, the ultimate goal in computer vision was the ability to identify objects within an image. Subsequently, the ability to discern the activities depicted in the image became a significant objective. If the 2012 ObjectNet competition marked a milestone in computer vision, then the remarkable achievements of transformers and language learning models can be equated to a similar event in natural language processing. These developments led to a decline in alternative approaches and a proliferation of similar initiatives.

1.1 Motivation

In the contemporary era, we witness remarkable advancements in understanding fundamental existential questions. These encompass the nature of the universe, its origin, the emergence of life, and the intricacies of intelligence. These profound questions appear to be on the cusp of elucidation. The motivation behind these inquiries is multifaceted. Perhaps it stems from the pursuit of knowledge that leads to practical applications for our survival. This drive for exploration is a central theme explored in the article entitled "Intrinsic Motivation for Artificial Agents," which is included in this collection.

A prevalent issue arises not from the inherent challenges of object recognition or the unreliability of LLMs, but rather from a situation where an excessive number of scientists are overly invested in the current impressive technologies, potentially hindering the exploration of other valuable scientific avenues. Is it truly necessary for everyone to follow the same path? Clearly, we do not think so. The recognition of the Geoff Hinton Turing Award is a testament to his unwavering commitment to his interests, even when they deviated from the prevailing trends.

1.2 Measurement

Whether defining intelligence in such a manner is worthwhile is debatable. Turing [6] circumvented the problem by saying that we would recognize intelligence when we encounter it. Although his approach has contributed to our current understanding, it also possesses flaws, as discussed in the article titled "Intelligence Testing for AI" in this volume. If we posit that intelligence is essential for playing a proficient game of chess and constructing a machine capable of achieving such proficiency, as IBM accomplished with Deep Blue [2], this does not necessarily imply that the machine possesses intelligence, even if it could. Conversely, we cannot assert that a machine incapable of playing a proficient game of chess lacks intelligence. Animals have often been claimed, especially historically, to lack intelligence and while they cannot play chess, numerous instances demonstrate learning and skill acquisition that many would characterize as evidence of intelligence [7].

When evaluating an AI against a human, it is crucial to consider not only one dimension but also the absence of specific capabilities, as this can diminish its usefulness in collaboration with humans. The paper included in this collection, titled "Intelligence Tests for AI," delves into the measurement of intelligence in humans and, more recently, in AI systems. In humans, intelligence tests serve as valuable tools for identifying individual weaknesses that can be addressed through training or career selection. The primary objective in humans is to enhance the success of the individual's life. Similarly, for AI, the goal is multifaceted, encompassing both the improvement of deficient areas and the overall improvement of the usefulness of AI systems. This approach may help prevent the tendency to prioritize the most successful dimension while neglecting the areas of deficiency.

1.3 Architecture

The human brain possesses the ability to comprehend images and evolving visual sequences. In particular, a relatively recent development is the ability to communicate with natural language. However, in this discussion, we refrain from naming specific brain regions. Planning and executing long-term objectives is another relatively recent addition to brain capabilities. Furthermore, the ability to learn from both mistakes and successes has proven instrumental in our survival and evolution. The paper titled "Society of LLMs: Constructivist learning approaches with multiple LLM instances," presented in this volume, proposes an architecture for constructing such a system involving multiple collaborating components.

It is crucial to recognize that the brain is a complex system composed of numerous interconnected components that collaborate to generate what we define as intelligent behavior. Intelligence is not solely determined by an algorithm; rather, it emerges from an architectural framework that embodies a comprehensive set of capabilities that constitute intelligence. If one or more of these capabilities are compromised, such as the ability to read and write, it is erroneous to assert that the individual lacks intelligence. In fact, the capacity to solve intricate problems and excel in a game of chess may persist. This does not imply that none of these dimensions is essential for intelligence; therefore, if all of them are diminished, the individual can still possess intelligence. These capabilities are valuable because they represent essential skills that contribute to a successful and fulfilling life.

1.4 Grounding

The topic of this year's workshop was 'Grounding' which refers to the process of connecting learned concepts with real-world meaning and sensory experiences. Grounding ensures that AI systems understand and interpret information in a way that is meaningful and relevant to the environment they interact with. The paper on *"Context-Centric Knowledge Representation (CCKR)"* in this collection proposes how such a grounding can occur. Grounding is essential for developing systems that truly understand and interact with the world. By linking abstract representations to real-world experiences, AI can demonstrate context-aware intelligence.

Grounding does not necessarily imply embodied intelligence, nor was embodiment the primary focus of the workshop, but it is an important consideration. The temptation to construct simple artificial worlds is difficult to resist, and recent remarkable demonstrations such as "hide and seek" [1], Checkers [4], Chess [2], and Go [5] are in various ways examples of how by simplifying the world, we can produce remarkably capable behavior. However, these approaches have significant shortcomings that cannot be easily remedied. They can be regarded as impasses in the pursuit of general AI.

1.5 Schemas

Embodiment is helpful in avoiding some of these shortfalls, in part because the real world is complicated, making it less easy to cheat! Humans learn to survive in a complex and often hostile world, and they can describe these approaches. Being able to describe them allows us to teach or demonstrate skills to others. Some of these skills may require complex fine motor skills that must be learned through practice. These kind of control systems, which may involve huge numbers of muscles and sensors, may well be best treated as neuro-inspired control systems,

At a higher level, representations will inevitably become symbolic. This is the neurosymbolic divide that the CCKR attempts to bridge. At a symbolic level, we can discuss picking up a saucer with the left hand and then placing the coffee cup on the saucer with the right hand. Such sequences of actions enable us

Fig. 2. This volume contains five papers that describe approaches to solving five research areas that may lead to AGI.

to share skills at the language level with individuals who have different kitchen layouts, muscle configurations, and touch sensors. However, these skills may eventually be represented as *schemas*. The paper titled "Schema Mechanisms 2.0 for Developmental Artificial Intelligence" presented in this volume provides an in-depth analysis of some existing schema implementations. By learning schemas at a symbolic level, it is possible to learn the skill before practicing the skill. If a skill has been learned on a single try, by being told, or by reading it in a book, the intrinsic motivation to practice skills may drive the agent to practice until the low-level bindings have been refined so that the skill can be executed with precision when it is appropriate. Researchers working at the schema level rarely address the refinement of skills through practice where that refinement is to a neuro-control system. Similarly, researchers in neuro control systems rarely deal with symbolic-level reasoning.

At the symbolic level, we can build plans that may contain contingencies. It is where we can add creativity, by handling the case where we don't have a saucer, but we can use a tray or a small plate instead of a saucer. It allows us to build a theory of mind that allows us to reason about what other people know and believe. Planning, Theory of Mind, and Creativity are all things that LLMs cannot do. We believe that because of this, it is essential that the neurocentric approaches and the symbolic approaches come together if we are to achieve AGI.

2 Conclusions

If the domain of human endeavors was similar to a game of golf, the 18th hole of artificial intelligence (AI) would be artificial general intelligence (AGI), as shown in Fig. 2. For individuals engaged in artificial developmental learning, AGI is not necessarily the immediate objective for most researchers, but we believe that our contributions are paving the way towards its realization. In this introductory section, our objective has been to organize the papers presented in this volume, all of which were collaborative endeavors, into a cohesive framework. The five papers collectively contribute to the five facets of the pentagon illustrated in Fig. 2.

From the top left, clockwise, the papers of the volume are as follows:

Grounding is covered in the first paper in the volume, titled *Concept-Centered Knowledge Representation: A 'Middle-Out' Approach Fusing the Symbolic-Subsymbolic Divide* addresses the challenge of bridging the gap between perception and understanding and between neuro and symbolic approaches. The authors claim that in addition to addressing the grounding problem, this will accelerate the progress towards AGI by bridging the neurosymbolic divide.

Architecture is addressed by the second paper, titled *Society of LLMs: Constructivist learning approaches with multiple LLM instances,* addresses the topic of architecture. It proposes a method for constructing a society of LLMs as a means of integrating disparate capabilities within an architecture that can provide complex problem-solving solutions. These solutions may encompass the ability to coordinate and leverage the capabilities of multiple LLMs, for example, the ability to plan over other capabilities.

Schemas are fundamental to developmental learning. The third paper in the volume *Schema Mechanisms 2.0 for Developmental Artificial Intelligence* provides a comprehensive review of existing schema-based learning systems. Schemas are acquired through interaction with the environment. For an agent, whether physical or virtual, to learn, it must interact with its environment, and this interaction is driven by the agent's motivations.

Motivation serves as the driving force behind action, and interactions with the environment foster the development of learned schemas. These schemas, in turn, enable the agent to respond effectively to the motivations driven by the environment. Motivations can be categorized as intrinsic, arising from an innate drive within the agent, such as the desire to master a skill through practice. In contrast, extrinsic motivations stem from explicit rewards, as exemplified by the pursuit of food in response to hunger. The fourth paper on *intrinsic motivation for artificial agents* systematically enumerates various artificial intrinsic motivations and explores their potential integration to generate novel and engaging behaviors that can subsequently enhance the schema learning capacity of the agent, enabling it to acquire useful skills in various situations.

Measurement serves as a crucial indicator of our progress. However, the significance of measurement lies in the evolving nature of intelligence testing. Currently, this field aims to assess the advanced capabilities of LLMs, yet it neglects dimensions where LLMs do not excel, such as planning, theory of mind, and comprehension of temporal sequences. The fifth paper, titled *Intelligence Tests for AI,* provides a comprehensive perspective on intelligence testing, encompassing both human and non-human intelligence. Ultimately, the nature of testing determines the outcomes we obtain; we get what we test for. Our objective should be to develop an AGI that demonstrates proficiency comparable to that of humans in domains deemed essential for human cognition. If we prioritize the creation of increasingly challenging mathematical tests while disregarding planning and other inherent weaknesses in LLMs, we risk developing a narrow-minded robot akin to savant syndrome. To achieve AGI, we must ensure that our testing methodologies encompass the full spectrum of cognitive abilities.

Acknowledgments. All acknowledgments, where appropriate, are in the individual papers introduced in this Introduction.

Disclosure of Interests. The authors have no competing interests.

References

1. Baker, B., et al.: Emergent tool use from multi-agent autocurricula. arXiv preprint arXiv:1909.07528 (2020)
2. Campbell, M., Hoane, A.J., Hsu, F.: Deep blue. Artif. Intell. **134**(1–2), 57–83 (2002)
3. Krizhevsky, A., Sutskever, I., Hinton, G.E.: Imagenet classification with deep convolutional neural networks. In: Advances in Neural Information Processing Systems (NeurIPS), vol. 25, pp. 1097–1105 (2012). https://papers.nips.cc/paper/2012/hash/c399862d3b9d6b76c8436e924a68c45b-Abstract.html
4. Samuel, A.L.: Some studies in machine learning using the game of checkers. IBM J. Res. Dev. **3**(3), 210–229 (1959)
5. Silver, D., et al.: Mastering the game of Go with deep neural networks and tree search. Nature **529**(7587), 484–489 (2016)
6. Turing, A.M.: Computing machinery and intelligence. Mind **59**(236), 433–460 (1950). http://www.jstor.org/stable/2251299
7. de Waal, F.: Are We Smart Enough to Know How Smart Animals Are? W. W. Norton & Company, New York (2016)

Concept-Centered Knowledge Representation: A 'Middle-Out' Approach Fusing the Symbolic-Subsymbolic Divide

Pei Wang[1] and Kristinn R. Thórisson[2,3](✉)

[1] Temple University, Philadelphia, PA, USA
pei.wang@temple.edu
[2] CADIA, Department of Computer Science, Reykjavik University, Reykjavik, Iceland
thorisson@ru.is
[3] Icelandic Institute for Intelligent Machines, Reykjavik, Iceland

Abstract. A long-standing paradigmatic debate in artificial intelligence is between the so-called *symbolic* and *connectionist* (or 'sub-symbolic') approaches to knowledge representation. Both approaches aim for uncovering the principles of how general, domain-independent knowledge is structured, generated, and handled by autonomous intelligent agents. Yet each seems to work only for certain kinds of information. The approaches spring from disjoint methodological beginnings; the former is inspired by human introspection ("top-down"), the latter by brain substrates ("bottom-up"). Neither approach has led to a unified theory of general intelligence. Few researchers are fluent in both methodologies, as using either approach calls for significant time and effort easily spanning decades. As a result, progress towards theories of general intelligence have been held hostage. We propose to brake this deadlock with a third approach: *Concept-Centered Knowledge Representation* (CCKR). Based around *situated dynamic knowledge graph generation and management*, CCKR captures latent features inherent in conceptual graphs that prior approaches do not address and adds capabilities that we argue are necessary for, and offer a path to, general machine intelligence. Here we explain CCKR and present arguments for its claims, resting in part on the results from two implemented experimental systems, the Non-Axiomatic Reasoning System (NARS) and the Autonomous Empirical Reasoning Architecture (AERA).

Keywords: knowledge representation · concept-centered representation · symbolic · sub-symbolic · connectionist · cognitive architecture · artificial intelligence · general intelligence · autonomy

1 The Topic of Representation

Knowledge representation has been a central topic of artificial intelligence (AI) from its beginnings. The main issue at stake is the question of how a controller

P. Robertson and O. Georgeon (Eds.): IWSSL 2024, LNCS 15457, pp. 10–52, 2025.
https://doi.org/10.1007/978-3-031-96325-4_2

can control a body[1] in a complex environment, for various purposes, where only a fraction of the environment (and itself) – variables, outcomes, solutions, goals, problems, etc. – can be known (observed, isolated, summarized, measured, etc.) at any moment. No matter what type of tasks or information is involved, an intelligent agent's knowledge about the domain it inhabits, and the problems and goals it faces, need to be available to be processed and used to take action. This calls for a *memory* and a *format* – a ***representation*** – for storing those memories as knowledge, that is, manipulable and editable information structures, to use for guiding future behavior and learning.

Given an aim of matching human cognition in numerous environments, an agent with artificial general intelligence (AGI) must ultimately be able to deal with the physical world, with its infinite potential for variety and variation. In contrast to engineering methods that carefully design both the environment and the controller/agent, as a couple, to achieve predefined ends using predefined means, here we are concerned with agents that are guaranteed to *not know* everything they need to know to get things done, and must thus *figure out stuff* for themselves [83] and *learn on the job*, to get better at setting and achieving goals. Furthermore, situations where a learning agent can only interact with (measure and affect) a fraction of the state space it finds itself in at any point in time – and throughout its lifetime – and its memory and processing power is too limited to keep all potentially relevant details in mind, even when known, and situations where it is impossible for an AI system's designers to provide a complete list of what should and must be known before the agent leaves the lab, are not just inevitable, they are the norm.

This is what defines our scope here: Autonomy, generality, and cumulative learning [85], in worlds as complex as the one humans live in. Why we should pick such a lofty and seemingly "impossible" scope is due simply to the fact that the vast majority of interesting problems, environments, tasks, and situations that we can think of – and for which we might want a *very smart* machine – all come with essentially those exact constraints. For this challenge, over-simplifying the requirements is an easy mistake to make, with the obvious risk of the research missing its mark, or in the very least delaying progress substantially.

Complex dynamic (non-random) worlds present vast amounts of information, and the role of learning is to systematically keep track of useful regularities[2] in a compact yet flexible format. It is the *form, formation*, and *use of information structures* resulting from such learning that is our focus here, in particular, with respect to attainment of *increased generality and autonomy* of the cognitive control mechanism—i.e. improvements in adaptation through informed learning (as

[1] From a control perspective, a controller's "body" is by definition demarcated by its sensorimotor operations and the substrate in which control processes are implemented, which define what it can in principle measure, affect, and know. For all practical purposes, the term can thus be interpreted widely – as in 'a controller plus its controlled plant' – or narrowly, as in 'a biological agent in its natural habitat.'

[2] That is, any regularly recurring patterns that can be reliably measured in the world (by the agent) and trusted for getting things done (by the agent) can be considered *useful to the agent.*

opposed to random experimentation), over a diverse set of tasks and environments. To do so, the knowledge representation must, in our view, directly address autonomous (a) re-formulation (selective improvement, editing), (b) strategic partial re-use, through compositionality, (e) incremental buildup and maintenance (cumulative learning), and (d) compaction/compression (selectively non-lossy and lossy), including strategic deletion (informed, selective forgetting).

We present a theory-guided [84,105] high-level *concept-centered knowledge representation* (CCKR) methodology for how the above requirements can be addressed. The approach aims to take even one step further, superseding prior attempts by offering a more unified, comprehensive, consolidating theory of representation for generally intelligent autonomous agents capable of *autonomous explainable cumulative learning.* The paper rests on a conceptual analysis, generalization, and discussion of representational approaches taken in two ongoing engineering projects with a combined six decades of research focused on autonomy and generality, the Autonomous Empirical Reasoning Architecture[3] (AERA;[4] [62,82,84,90]) and the Non-Axiomatic Reasoning System (NARS[5] [93,94,97,103,106]).

CCKR can be seen simultaneously as a theory of concepts and a theory of how to implement such mechanisms in a machine [106]. The approach rejects the top-down methodology of the symbolic stance and the draconian bottom-up approach of neural-based approaches; CCKR can be thought of as a 'middle-out'[6] approach that starts from the ingredients that a thinking mind uses to create, manipulate and manage concepts, where the "upwards" direction from this middle ground links to higher-level phenomena such as plans, goals, and meaning, and the "downward" direction links to whichever substrate the mind is implemented—be it neurons, silicon, or something else entirely.

This is our first systematic consolidation and presentation of a CCKR approach, and the first exposition and analysis of how it challenges, and goes beyond, strictly symbolic and connectionist approaches to knowledge representation. While sharing some similarities with more common schools of thought, CCKR is in many important aspects fundamentally different from most well-known approaches to knowledge representation, as our demonstrations here of its principles show.

Several distinct meanings of the term *concept* can be identified in everyday language. A prominent one is as a synonym or placeholder for a "category" of everyday phenomena, e.g. "the concept of *a chair*," "the concept of *fast*." Another (and closely related) role is for referencing immaterial ideas (for instance, mathematical concepts like 'perfect circle' and 'infinity'). On an intuitive level, our use of the term can be said to be fairly compatible with many such uses, but since our aim is to provide a more rigorous meaning to it than a standard dictionary

[3] Also called the 'Autocatalytic Endogenous Reflective Architecture' [60].

[4] http://www.openaera.org – *accessed Jan. 7th, 2025.*

[5] http://www.opennars.org – *accessed Jan. 7th, 2025.*

[6] Thanks to Mike Judge and his co-authors of Silicon Valley for proposing this concept.

definition, some correspondences with the vernacular are inevitably weakened or abandoned.

1.1 Scope and Organization

After a short overview of selected methodological approaches to knowledge representation, we review and analyze the main features of the so-called 'symbolic' and 'connectionist' schools, and prior attempts to combine them. Then our *concept-centered approach to knowledge representation* (CCKR) is presented. To show the realizations and implications of CCKR, NARS [103] and AERA [62] are described, compared, and contrasted. Both systems are based on a CCKR approach and rest on common foundational assumptions about intelligence and cognition, while their differences expose some of the space in which arguments about further details, assumptions, and specifications of the CCKR framework must play out. To stay focused on arguments backed up by evidence, we limit any claims, descriptions, and examples of CCKR to the *existing implementations* of these two systems.

We also look at arguments for how CCKR subsumes prior approaches and other conceptualizations of knowledge representation, and compare our approach to existing ones, highlighting the advantages of CCKR in the development of generally intelligent systems. Lastly, we discuss the open issues. As our focus in this paper is squarely on representational topics, the discussion will center first and foremost on issues considered necessary for that, rather than on how the representation may be used in various cognitive functions of AGI systems.

2 Knowledge Representation: Early History

The numerous approaches to knowledge representation explored in artificial intelligence (AI) have been influenced by several other disciplines including computer science, psychology, mathematics and logic, linguistics, philosophy, and neuroscience; here we will limit our overview to the first three.

2.1 The Road to Representation

Work on knowledge representation grew out of early AI research and the study of data structures in computer science (c.f. [53,67]). The subject of representation is familiar to any computer programmer, as computers cannot do any computing without representing information in some way. Going well beyond common structures such as sets, lists, trees, and graphs,[7] AI calls for representations that support learning and reasoning. Artificial general intelligence (AGI) makes additional requirements, including that the learning extends to a wide range of topics,

[7] While many information structures can be transformed into each other, here we are not concerned with structural and implementation details but rather operational properties relevant to the conceptual design of their use for understanding cognition.

tasks, situations, and environments, and that the learner be capable of imbuing this acquired knowledge with meaning (for itself, its tasks, its owner, and/or its environment), making it increasingly autonomous. The higher one climbs up the "ladder of intelligence," towards increased generality and autonomy, the stricter such requirements seem to get. While the full list of requirements for AGI is still being debated [88], it seems likely to require representation schemes that go far beyond the well-defined expressiveness, versatility, automated organization and processing efficiency of traditional programming languages, implying self-generated meaningfulness and flexible modularity in support of that self-organization [59,86,89].

Early work by the pioneers of the field focused intently on representation as a key aspect of higher-level cognition. However, progress in AI proceeded slowly in the first few decades, made only more obvious in light of the frequent overoptimistic predictions of imminent near-term progress. Results from research on humans in cognitive science seemed only to confuse matters, in part due to a misalignment of short- and intermediate-term objectives [108].

2.2 Behaviorist Approaches

In both psychology and AI, attempts were made to outlaw representation and focus instead on the world an intelligent agent inhabits [12,79,109]. To this end, Brooks proposed "intelligence without representation," suggesting that representations were, in fact, wholly unnecessary, as "[i]t turns out to be better to use the world as its own model" [12, p. 140]. In support of this extreme theoretical stance, he proposed a method of networked augmented finite state machines (AFSMs) for constructing robot control mechanisms for (semi-structured[8]) physical environments [11,13]. However, instead of being representation-free, as the intent and claim of that research was, subsumption-based systems in fact *baked in* the representations directly from the very beginning, in the AFSMs network structures themselves: An AFSM-based controller necessarily includes hard-wired sensing, acting, and goal structures that rely on (the designer's) assumptions about the agent's environment, and the agent's future interactions with it. (Perhaps a more appropriate name for would have been 'Assumption Architecture'). While the subsumption approach does indeed assume, unlike behaviorism, the existence of internal goal structures, these generally cannot change automatically through learning or adaptation after the systems are deployed, due to features inherent in the methodology, and the approach thus not only fails to meet one of the key requirements for (general) intelligence, i.e. the ability to change one's mind – to re-evaluate goals, derive new goals, and to abandon them in light of new evidence [83,102] – it also fails to achieve what it set out to do, namely, to do away with "internal representation."

Like the behaviorist psychology movement [79,109], the subsumption architecture shifted the representational discussion in AI research to the environment,

[8] The methodology was thoroughly tested in household robots, most famously the vacuum cleaners of iRobot, a company Brooks co-founded.

in the hope of simplifying the research related to the control system at its center, attempting – but inevitably failing – to free its very subject matter from the theoretical basis on which it in fact depended. No behaviorist-themed approach has succeeded, or in fact will ever succeed, in ostracizing representations from an agent whose behavior can be considered in any way goal-oriented: The issue will end up being a blind spot in the methodology, or the representations given an implicit and entangled incarnation in the control system's very design. The cost is a theoretical handicap and cognitive inflexibility of the resulting systems.

2.3 'Subsymbolic' Approaches

Early interest in engineered systems inspired by natural neural networks dates as far back as the early cybernetics research of Wiener et al. (cf. [67]). Historical highlights of the development of ANNs (in a broad sense) include parallel distributed processing (PDP) "connectionist" models [71, 74, 80], Hierarchical Temporal Memory (HTM; [32]) and most recently, Deep Learning [30, 44, 75].[9]

Minsky's Stochastic Neural Analog Reinforcement Calculator (SNARC [52]), and Rosenblatt's Perceptron [72], were the first demonstrators of how to implement what came to be called 'artificial neural networks' (ANNs). Minsky, however, soon abandoned the approach entirely, along with another founding father of the field, John McCarthy [48], on the grounds that it wasn't sufficiently explicit, preferring strictly symbolic approaches (c.f. footnote 16 on page 21).

A variant of connectionist approaches is the so-called *dynamical hypothesis* of cognition [8, 24] (see further discussion on its roots on page 23). Influenced by research on chaotic systems, it borrows numerous abstract concepts from physics such as attractors, phase-space, and limit-sets. We consider its philosophical stance to be relevant here for at least two reasons. Firstly, on the positive side, more than the other schools of thought, it makes *time* a first-class citizen of cognition, something we consider a necessary (but not sufficient) requirement for research on general intelligence. Secondly, and on the negative side, by borrowing concepts invented to explain the behavior of low-level non-cognitive physical processes, and positioning these as part of a serious proto-theory of (very) high-level properties of intelligent systems (mental operations and cognitive control), its conceptual underpinnings sorely lack an intermediate level of explanation and operation. While being relegated to a descriptive, narrative or inspirational role at best, a significant level gap nevertheless exists between its theoretical concepts and the phenomena it purports to explain. As a result, dynamical approaches are insufficient for guiding the construction of complex cognitive systems in any specific way, and cannot provide a viable AGI research methodology. This shortcoming is shared by a number of other approaches, including behaviorism and neurology. In the case of both neuroscience and dynamical approaches, we

[9] These approaches are 'sub'-symbolic in that they generally focus on information at smaller scales than, or schemes that are orthogonal to, concepts (in the vernacular meaning); probably the best known of such approaches, 'connectionist' ones depend on a distributed (fragmented) representation.

estimate this spatiotemporal gap that they fail to address to span several orders of magnitude. It is in part this gap that CCKR fills.

2.4 The Need for Explicit Representation

Like the founders of AI research, we consider representation to be central research topic for the field. For any research program addressing general intelligence directly, representations cannot be left out, and nowhere is the argument for this made more clearly than in Conant and Ashby's *Good Regulator Theorem*, where the role of models is explicated on solid mathematical grounds[10] [15]. Their work makes clear that it is not only desirable but in fact *impossible* to create a working controller without representation.

3 Representation: Symbolic vs. Connectionist

In this section we review the symbolic–connectionist debate from our perspective and set the context for how our approach is an alternative to the two.

3.1 Connectionist Representation

The connectionist approach comes from the idea that intelligence can be obtained by building a brain-like neural network – an idea dating back more than 100 years – and has been fueled by the recent success in applying artificial neural networks (ANNs) to a wide variety of information-centric tasks. Ever since the cyberneticists' work on (natural) neural networks [50], the model followed has been based on a large number of simple units connecting to each other via weights that are tuned (during "training") using special algorithms. Besides the obvious (albeit negligible) inspiration from neuroscience, this approach has been heavily influenced by mathematics research.

The typical form of connectionist AI is an ANN that is trained over numerous iterations on a given dataset, during which the weights of the connections are tuned. After an ANN is trained, the network's input/output layers forms a mapping for commonalities extracted from the training data, while the operation of the intermediate (hidden) layers is determined by the role needed to support this overall input-output mapping, i.e., the function represented by the network. Consequently, a vector that forms an input or output layer represents a sensation pattern (input) or decision (output), while a vector from a hidden layer of the ANN does not explicitly represent anything outside the network. Instead,

[10] Models come in infinite forms for an infinite variety of purposes. In a learning controller they are *empirical* when they are based on *empirical experience* (that is, in a way that can be argued to be based in some direct or indirecty way on measurement—i.e. 'grounded.'). In Conant and Ashby's theorem they could be formal ones [15], but need not necessarily be, and in fact cannot be when the axioms are not known (in which case they are empirical). Unless otherwise noted, it is these kinds of models we mean when we use this term.

it contributes to overall mapping of the model in a way that, for a particular ANN, is difficult to explain in succinct and simple terms. For this reason, ANNs excel as "pre-cooked" automatic classifiers; when posing as controllers, however, their behavior is difficult to ensure and predict, especially on data at the fringes of, and outside, their training data.

Though the various types of ANNs differ in many aspects, they share the use of what is called a *distributed* knowledge representation, which is fundamentally different from the *local* representation used in most symbolic AI systems. In distributed representation, "[e]ach entity is represented by a pattern of activity distributed over many computing elements, and each computing element is involved in representing many different entities" [34, p. 77]. In contrast, in the symbolic tradition, each symbol can only have a single specific reference at any time.

To make use of an ANN in a practical setting, a fully trained ANN is typically placed in a larger architecture where it performs the classifying function it was trained for, serving as the representational component of a control system made up of other components. While most would say that the ANN "holds" particular representations that allow it to perform a complex function that requires "knowledge," this representation is unchanging and unchangeable in light of new evidence or information, until it is re-trained by its designers. Insofar as the "knowledge"[11] of the ANN references objects and phenomena outside itself, the output layer can be seen to implement a symbol system. However, this implies that the input-output mapping performed by an ANN – its 'knowledge-based' classification function – is grounded in the outside environment only at its input and output *ends* (in a fixed manner), as specified by the function. The ANN does not generate or obtain such knowledge itself, autonomously, and this information has thus no meaning to the ANN, only to its designers. Because the representation that the ANN holds is allonomically determined, i.e. decided, defined, and interpreted by the engineers during input-output training, it has been engineered to mirror an input-output mapping in their (human) knowledge, in light of *their* goals—not the ANN's. Such knowledge cannot be re-assessed on the job, so to speak, by the system itself—its reliable modification, no matter how small, requires a 'recall and full reset' of the system as a whole.

Generally speaking, connectionist approaches favor data quantity and representational uniformity over representational explicitness. The resulting knowledge lacks structural compositionality (modularity), inspectability, and explainability, to name three serious limitations in the pursuit of general intelligence. This, along with the features described in the foregoing paragraph, is the main reason why ANNs are always embedded in a larger control architec-

[11] We consider 'knowledge' to be 'information that can be used to compose actions in pursuit of an explicit or implicit goal'. While an ANN certainly holds information that can be used to act upon, the ways in which an ANN can act on its knowledge is highly limited (note that this is different from the scope of *input* that an ANN can accept – which is limited in other ways, too complex and removed from the purposes of this paper to discuss here).

ture: It does not "know what to do with its own knowledge" beyond the function for which it has been designed, as it lacks any facility to provide meaning to its internal information units independent of this function. Furthermore, intermediate nodes and node groups (e.g. layers) in an ANN may or may not map to anything in particular outside the training data and/or in other knowledge sets (such as its human engineers' knowledge). The intermediate outputs of sub-groups of nodes in the ANN are therefore meaningless (and largely unpredictable) from the human users' – as well as its own – perspectives, and one reason why the runtime operation of an ANN is difficult to understand.[12]

In summary, in consideration of the topic of this paper, the connectionist approach of knowledge representation is characterized by the following properties:

1. The system's knowledge is represented as a network of connected units which is often considered as approximations of neurons in a brain.
2. The system implements a function that maps the status of the input units (as a vector imposed by the environment) to that of the output units (also a vector as the system's response to the environment).
3. The input and output units get their meaning from their correspondence to (sensory, verbal, or other kind of) stimuli and responses, while the other (internal or hidden) units have no external correspondence and their roles are revealed by their contributions to the overall function.
4. Many human concepts may correspond approximately to status (activations) patterns of the network, which is *distributed* among the units and connections, rather than *local* to a single unit or connection.

Compared to the alternative approaches, the connectionist approach has nevertheless several major advantages:

- The function can be learned from training data in a connectionist network with universal approximation power, and does not require detailed human design.
- The learning process has a certain level of tolerance to uncertainty in the training data.
- The knowledge needed to build the function does not require a precise description.

Despite its notable achievements, the connectionist approach has met many fundamental objections over the decades, most of which are still unanswered by today's ANN methodologies and implementations. We can summarize the major criticisms on the representational aspect as such:

[12] The end-to-end usage of a neural network still has the symbol grounding problem at its input/output layers, except when the data directly come from sensors or go to actuators. A hidden layer has no such problem, not because it has grounded meaning, but because it has no identifiable *meaning* to talk about at all.

- The distributed representation in ANNs is very different from how human knowledge is expressed, which is one reason why ANNs have an explainability issue and adversarial examples [57], as well as difficulty of receiving human knowledge [68].
- Since knowledge is coded as numbers with a single kind of relation ("strength"), inspection on particular knowledge is not supported [54], which also leads to problems in local revision and cumulative learning [87].
- Since knowledge eventually takes the form of an end-to-end mapping, there is no identifiable components that recursively form constructions. In general, it means that ANNs still lack compositionality, hierarchy and systematicity as criticized decades ago [22,82].
- In input/output mappings, there is no distinction between correlation and causation [65,91].

No doubt, the listed objections differs in severity and it may be argued that solutions to some of the above objections, which date back to 1986 or even earlier, are already at hand. Indeed, some *ad hoc* solutions exists for special cases of the above limitations. Every such solution, however, is only partial and comes with severe limitations. Most scathingly for research in general intelligence, there exists *no solution that addresses all of them together*, or even some notable proportion, and this is by far the most serious limitation of all such work to date.

3.2 Symbolic Representation

Historically, what has been called a "symbolic approach" to representation can be traced to a focus on human cognition. Meeting a need to constrain the discussion of human cognition in some way, a narrow focus on board games centered on the widely-articulated hypothesis that the game of chess could only be successfully performed by a general intelligence.[13] Looking at the history of AI, this hypothesis could not have been more wrong. The hypothesis and approach comes from the (arguably somewhat intuitively compelling) idea that AI needs human-like knowledge representation to solve complex problems. For the most part it meant that symbols were used to represent external objects and events, which were then to be manipulated via rule-based symbolic processes. The approach, also known as GOFAI ("good old-fashioned artificial intelligence") [31], rests on a tradition tracing back to the study of logic [40], psychology [66], linguistics [14], and philosophy [21].

[13] The claim of Russian mathematician Alexander Kronrod in 1965, that chess 'is the drosophilia of AI' [20], may have sealed this particular board game as a major measuring stick on the advancement of AI, and quite possibly the focus on board games in general. Another reason for the field's obsession with chess was the hypothesis of Carnegie-Mellon researchers Newell and Simon [56] that a machine capable of winning a human world champion in chess would inevitably, by default, have obtained general intelligence, because that is required for playing chess—a hypothesis that could hardly have been more incorrect.

In AI, one of the best known incarnation of this view is Newell and Simon's *physical symbol system hypothesis*: "A physical symbol system has the necessary and sufficient means for general intelligent action." [56, p. 116].[14] It uses the concept of a "symbol" within a system to represent an "object" or "relation" outside the system, with the two connected by designation and interpretation. With such a representation, human knowledge in a domain, such as physics or biology, could be used by a computing system [33].

In this tradition, the symbolic representation was largely implemented in some sort of practical programmable logic, including first-order logic and extensions thereof, and resulted in a fundamental emphasis on formal logic and related automatic reasoning. In this camp was John McCarthy's theory of intelligence, according to which people used 'commonsense reasoning' to solve problems [48]. This theoretical foundation resulted in several research programs that attempted to mechanize expert knowledge, since symbolic knowledge encoded as logic statements, coupled with the proper (hand-coded) heuristics, should suffice to mechanize thought in a particular domain, allowing a machine to take the role of a domain expert through the systematic manipulation of symbols. The scope this approach was then extended to all human knowledge, leading among other things to the Cyc "common sense database" [45], which remains one of the largest research programs in AI to date. While the work by Lenat and Cycorp[15] on the Cyc system is unique in the history of AI, it had the support of Minsky, who, like McCarthy, was a strong proponent of a symbolic AI and dismissive of all connectionist approaches [54].[16]

For the current discussion, the basics of this approach can be summarized as the following:

- A 'symbol' is a token, sign or pattern with no particular or intrinsic meaning.
- A symbol is a 'stand-in' or "pointer" to something else.
- Anything can serve as a symbol of anything else.
- A system can use a symbol to represent an arbitrary outside object or event.
- The symbol and what it represents are related by an interpretation.
- The processing of a symbol by the system is fully determined by its shape, independent of its interpretation.

[14] The full paragraph in question states: "*A physical symbol system has the necessary and sufficient means for general intelligent action. By 'necessary' we mean that any system that exhibits general intelligence will prove upon analysis to be a physical symbol system. By 'sufficient' we mean that any physical symbol system of sufficient size can be organized further to exhibit general intelligence. By 'general intelligent action' we wish to indicate the same scope of intelligence as we see in human action: that in any real situation behavior appropriate to the ends of the system and adaptive to the demands of the environment can occur, within some limits of speed and complexity*".

[15] http://www.cycorp.com – *accessed on Apr. 12th, 2023.*

[16] In his keynote address at the 2005 annual conference of AAAI, attended by one of us (KRTh), Minsky stated that "statistical approaches to AI offer nothing under the hood to work with," preventing, among other things, reflective cognition. This view is echoed in his 2006 book *The Emotion Machine* [54].

It is often neglected that this usage of the word "symbol" does *not* exclude some binary strings in a computer system: If a binary string represents nothing in particular, or something that is not an outside object of event, then it is not a "symbol" in the above sense. In other words, a symbol, according to this definition, must symbolize an *external* entity or phenomenon, and can be seen as the name or label of something that exist outside of the system holding the symbol. It is only in this sense that it could be argued that there exist "non-symbolic" AI systems – even if they use binary strings or variable names in their implementation that do not "represent" or "symbolize" any outside object or event.

A major attraction of the symbolic approach to AI researchers is its similarity to how human knowledge is often represented in everyday speeches and writings, as well as mathematics, etc., and its apparent closeness to the knowledge-handling traditions of related disciplines such as psychology, linguistics, and logic.

In view of the goal of AI to create a general learner, this approach to knowledge representation has the following features:

Modularity:

- Relatively fine-grained knowledge representation (consisting of small tokens).
- Any token can act as a symbol.
- Can be used to compose larger units according to a small finite set of compositional rules.
- Allows modular construction and de-construction of knowledge.

The above makes the knowledge representation explicit:

Transparency:

- The representation format is relatively explainable and understandable (by a system's human developers).
- The content is inspectable at a fine level of granularity.
- The meaning of the representation is systematic and compositional.

Although the symbolic approach dominated AI research for decades in various forms, it has been criticized from a variety of philosophical, psychological, and pragmatic angles, including that:

- The symbolic approach ignores the dependency of knowledge on an agent's body and context [4, 25, 42]. A symbol is meaningless until it is grounded in something other than other meaningless symbols [29, 76].
- The approach is too rigid and brittle, and fails to capture the fluidity and creativity of the human mind [26, 37, 39]. Human knowledge in many domains cannot be fully formalized, abstracted, or symbolized [18].
- Its processing is limited to deduction on abstract knowledge, and is far from covering practical human knowledge [10, 51], especially procedural knowledge [110].

- To manually code all relevant human knowledge into AI systems is unrealistic and inefficient [82].
- The representation languages developed in the GOFAI tradition do not properly address how the knowledge could be assembled autonomously by AI systems [84].

3.3 Variants and Hybrids

Besides typical symbolic and connectionist approaches in knowledge representation, there exist approaches that are hybrids, as well as variants, of the two. One example that could be considered a combination of both is the above-mentioned *dynamical hypothesis* of cognition [8,24]. Dynamic-system representation sees the system's state as a point in a multidimensional space, and the running process as a trajectory in that space. In this way, it is similar to a connectionist approach. On the other hand, when each dimension is considered as a specific measurement, it also shares properties with symbolic approaches.

Another approach that is between the two paradigms is the *Slipnet* of the *Copycat* system [36], a network of fluid concepts that are neither symbols denoting external objects nor models of neurons that work as functions.

Many attempts have been made to integrate symbolic and connectionist representations in one system [9,16,47]. One school combines the two approaches as cognitive modules in an integrated architecture, such as CLARION [81], ACT-R [1], SOAR [41], and Sigma [73]. Another school explores "how principles of symbolic computation can be implemented by connectionist mechanisms and how subsymbolic computation can be described and analysed in logical terms" [9, p. 3] using "neurosymbolic" systems [17,23]. Concrete ideas for how to do this includes for instance using a neural network with dynamic external memory [27] and combining deep learning with Cyc [47].

These attempts at reconciliation invariably present some interesting ideas and demonstrate well-defined benefits over and above the approaches they combine, but they are, without exception, limited to sub-problems or special cases of general cognition. The bigger question of architectural unification and extension up to human-level cognition remains largely unaddressed.

4 CCKR: Concept-Centered Knowledge Representation

We now introduce an approach to knowledge representation we refer to as *concept-centered knowledge representation* (CCKR). While compatible with many high-level conceptualization of concepts in cognitive science (cf. [5]), our approach is based on concretely implementable ideas for how to build machines with a capacity for dynamic concept creation and use, cumulative learning, and empirical reasoning—in short, autonomous cognitive operation. To make the descriptions concrete, two AGI-aspiring systems we developed separately, NARS (Non-Axiomatic Reasoning System) [93,97,103] and AERA (Autonomous

Empirical Reasoning Architecture)[17] [61,62,82], are used as examples of how CCKR can be realized.

Intuitively speaking, at the highest level of abstraction, this approach takes the knowledge repository (memory) of a knowledge-based control system to be a conceptual (latent) network or graph, where nodes are concepts and edges are conceptual relations. To be more specific, the core ideas are:

1. Each *concept* is an identifiable (but flexible) information structure within a system that recaps (models) a portion of the system's *experience*.
2. The *general meaning* of a concept to the system is determined by its (experienced or imagined) relations to other concepts. Each time a concept is used, usually only a small part of its general meaning is involved, forming the concept's *immediate, spatio-temporally-bound meaning*.[a]
3. New concepts and conceptual relations are *constructed* by the system itself, to better model its experience.

[a] This refers to a concept's *foundational meaning* to the concept's autonomous owner; for a theoretical definition, see Thórisson & Talevi [92].

Further unique features define our proposal, of course, but these are the most important ones at the highest level. We can now proceed to explain each of the three ideas one by one, each in a subsection.

4.1 The Notion of 'Concept' in CCKR

Like all fundamental notions, "concept" has many usages and interpretations [43]. Our usage in CCKR is as follows:

Definition of 'Concept' in CCKR

We consider any data structure or information item a ***concept*** if – for any cognitive system capable of creating new knowledge (i.e., actionable information), that systematically relates to its goals, existing knowledge, and situation – the information item:

1. Is a unit that can be uniquely recognized/isolated (by a cognitive system's processes) and
2. one that both abstractly and concretely summaries certain segments of the system's experience, and

[17] Also called the 'Autocatalytic Endogenous Reflective Architecture' [60].

3. can be accessed and manipulated (by a cognitive system's processes) for particular purposes and operations (overt and covert goals), and
4. can (uniquely, fully, or partially) be related to other comparable or different units (by cognitive processes operating on them).

Data about the history of manipulation and use of the above information structures is embedded with it, so as to help consolidate, group, associate, and access, the information in the future. In this view, a concept can be seen as any information on the path to becoming, or already being, a relatively stable subset of a knowledge base that serves a practical purpose in the cognitive system's operations. Concepts are thus both macro- and microsymbolic[19] entities that not fixed, but rather, information sets that are coupled, to varying degrees of *strength*, between themselves and other such sets; a set whose elements are loosely-coupled holds a relatively vague concept, a tightly-coupled one is "crisp" or "clear." By being experience-centered and initiated, knowledge structures of this kind are not very much like traditional semantic networks in any way but rather an ever-changing network of dynamically-coupled augmented-symbolic information structures that are constantly being honed through experience of their use.

Concepts are thus chiseled by interaction with an environment and a social cohort, and thus take some time to form; fleeting and contrived concepts are possible – e.g. 'winged miniature basketballs' – as are strong, immutable ones—'mother,' 'cat,' and 'wind,' to take some examples from human experience that would be predicted by CCKR. Concepts exist because a cognitive mind has use for them; they become entrenched because they prove their value and get repeated use over long periods of time, which also enables them to have richer relations to other ones, both due to similarities and dissimilarities. Due to their compositionality (i.e. being hierarchical compositions of smaller concepts and knowledge elements), and the need for adaptation to a particular situation (which often involves a number of unique elements, e.g. a new goal that an agent has never pursued before), they are created or decomposed on-demand by appropriate cognitive operations in situ.

According to our usage, a concept cannot be a "nameless"[20] process or event that simply runs its course within the agent, whose side effects are untraceable after they happen (e.g. like an activation pattern in a neural network that is not

[19] In compositional representational approaches, we use the term 'microsymbolic' to refer to *items* that are *parts of* symbols in such systems, yet can also stand on their own and be accessible to the system through reflection; conversely, 'macrosymbolic' refers to concepts that rely on and/or encapsulate other concepts.

[20] It cannot be 'nameless' in the sense that it must be recognizable, manipulable, and operationable as a *unit*. here are undoubtedly many ways in which this requirement may be achieved, so this does not mean that we must resort to GOFAI or naïvely symbolic approaches, since in this discussion a 'name' is not a 'symbol,' but more of a 'pointer.' This will become clearer in the following sections; see also Sect. 6.2, page 41.

associated with a name or an identifiable trace). Nevertheless, the concrete form and content of a concept can still be different in different systems, and different at different times, under these requirements.

NARS uses a formal language, *Narsese*, to represent its knowledge. The basic unit of the language is *term*, and each term normally refers to a *concept*, which is a data structure in the system. In the simplest form, a Narsese term is a string of characters from a given alphabet. If the term appears in the system's input and output communication, it corresponds to a convention followed by the systems participating the communication, and the same term provides a way (at least partially and temporarily) to associate the concepts in different systems. Similarly, a term can be a word or phrase in a human language.

NARS also has concepts associated with sensorimotor devices and operations. Designed as a general-purpose system, NARS is not equipped with a specific set of sensors and actuators, but allows any hardware or software to be registered at a sensorimotor interface of the system. Each registered device, as well as each executable command of the device, will be taken as a concept, with a unique term as its identifier. In this way, NARS can issue commands to a device to carry out an action and get results and feedback.

Narsese uses a set of term constructors that each can build compound terms of a certain type from the existing terms, recursively. A compound term, including a compound perceptual terms and an operational terms, also refers to a concept.

AERA takes a somewhat different approach to CCKR than NARS. Among its fundamental atomic information structures relevant here are *variables*, *composite states*, and *models* (composite and atomic).[21] Replicode, the language an AREA agent's knowledge is represented in [59], is late-binding throughout, and variables work both as placeholders, pointers, values, lists, and arrays; composite states are variable combinations, and causal-relational models (CRMs) are a kind of bi-directional copula that may represent any transformation or relation, an important type of which is, as the name implies, causal [62,91]. CRMs have a left-hand side (LHS) and a right-hand side (RHS); to represent a transformation, the LHS is a prior state (typically represented with a composite state) and the RHS is the result of a transformation. The model holds the function for the transformation it describes (how the values on the LHS change to those on the RHS), along with the context that the model is relevant for.

A single CRM can itself be considered as a concept according to our definition, albeit a peewee one, from which other concepts are constructed. Most concepts in AERA are composed of a large number of models, and the complete set of such models is not fixed, as they are in part contextually defined. Their operational semantics emerge during the system's experience with using them in context, which ensures their grounding and forms the basis for the system's learning. For instance, the concept of a deictic gesture (pointing at something,

[21] We restrict our discussion here to information structures that an AERA agent uses for autonomous learning; other information structures can be used as part of a programmer-defined *seed*, but this is a more involved subject that need not be address here.

e.g. with your finger) would – depending on an agent's experience – include several models describing the shape of the hand, the relation between the object pointed to and the finger/hand, as well as how to time the occurrence of such a gesture with respect to speech (see e.g. [90]). The concept would not necessarily have any name associated with it, but rather would be summoned in part or in whole, as needed, through relevance mechanisms, depending on its intended usage in the current context (e.g. whether for generating a pointing gesture or for recognizing it when performed by someone else).

The knowledge that an experienced AERA agent has learned about a complex domain from the ground up will consist of a very large set of such peewee models [58]—the size of the set being a function of how knowledgeable the agent is.[22] The largest such knowledge base created from scratch through an agent's experience, as demonstrated to date, was our S1 agent that learned to conduct a spoken-language multimodal interview with a human by watching humans do it. This agent's knowledge started with 26 hand-coded bootstrapping models; after learning how to perform an interview it contained 1400 models [90], which it had created autonomously over a course of 20 h of observing two people in realtime dialog.

CRMs can effectively encode causal relations ("affordances") such as "sittable-on," which allows them to be used effectively for plans and action. The concept of 'chair,' for instance, could have dozens, hundreds, or even thousands of AERA models—many of which would be shared with other concepts (CRMs natively support transfer learning mechanisms in AERA), but some of which would be unique to chairs. The couple of "sittable-on" and "single-person," each of which could be represented by a single CRM (but will then reference other models as well, mostly lower-level), will *together* be rather unique to chairs, even though sofas, benches and the side-walk can also be sat on (and thus considered 'seats' but not 'chairs'[23]).

4.2 The Meaning of a Concept in CCKR

According to CCKR, the bulk of a learning agent's memory, at a particular point in time, consists of a (implicit or explicit) concept network that as a whole contains the systems' total knowledge, and which, given the set of possible

[22] Our knowledge representation method allows re-use of knowledge, considerably reducing representational needs for domains rich in self-similarity (in the limit, memory increase for learning an additional domain $\mathcal{D}_a$ at a single level of detail, $f_{mem\uparrow}(\mathcal{D}_a)$, is close to the memory requirements for additionally storing the difference between the new domain and the most similar one already learned, at that level of detail, $\mathcal{D}$, assuming perception/acquisition of their differences, $\Delta = \mathcal{P}(\mathcal{D} \setminus \mathcal{D}_a)$, approaches perfection: $\lim_{\Delta \to \infty} f_{mem\uparrow}(\mathcal{D}_a) = \mathcal{D}_a - \mathcal{D} + \alpha$, where α is a small fractional overhead).

[23] It should be noted that here we are using these words for human concepts purely for convenience's sake; an AERA agent that learns concepts from scratch will be entirely limited to its own world of experience and may thus not end up holding such human-centric concepts at all.

operations on them, defines the limits of everything that the systems can be said to "know" at the moment.[24] In this network, the nodes are concepts, and the links are conceptual relations. There is a number of basic link-types, each operating in accordance with particular semantically closed operations [64], that is, what such a link means is defined completely by what the system can do with the link, in accordance with our definition above (see p. 24). The other (non-basic) relations between concepts are concepts themselves, and they are related to the basic types.[25] The network changes constantly (in terms of its topological structure as well as the attributes of the nodes and links) when the system is running.

Given a conceptual network of this kind, the meaning of a concept to the system is determined by its relations with other concepts, and these relations have (almost exclusively) come from the system's *experience*, which consists of spatially localized streams of input and output over time. Each such stream consists of conceptual relations itself, including temporal and spatial relations as special cases. This way, the meaning of a concept in CCKR is "experience-grounded," and is accessible to the system, rather than given by the system's designer through pre-conceived notions, or by an interpretation that maps it to outside entities.

In NARS, the basic conceptual relations are the *copulas* that intuitively link two terms into a statement. There are four of them:

Inheritance: S → P means that "S is a type of P."
Similarity: S ↔ P means that "S is similar to P."
Implication: C ⇒ R means that "C implies R."
Equivalence: C ⇔ R means that "C is equivalent to R."

The accurate operational semantics of these copulas are provided by the inference rules of NAL [103]; the first two specify when a term can replace another one by meaning, while the last two specify when a statement can replace another one by truth-value. Since terms (including statements as a special type) name concepts, reasoning in NARS is effectively all about the substitutability between concepts and the propagation of such substitutability.

Using the copulas as basic built-in relations, other relations can be defined. For instance, an arbitrary relation R among arguments A, B, and C can be represented in NARS as (*, A, B, C) → R where the left side of the arrow is a compound term intuitively similar to the Cartesian product defined in set theory, and this statement literally states that the relation among A, B, and C is a special type of the relation R. In this way, the inference rules of NARS only

[24] In AERA, the network is implicit and extremely large, so should not be processed as a graph. In NARS, the network is explicit and can also be huge, though it will never be processed as a whole. In both systems the networks are always processed partly and incrementally. The memory of NARS may look like a semantic network or knowledge graph, yet there are fundamental differences (as explained below).

[25] A concrete example is given latter, where relation R among arguments A, B, and C is represented in NARS as (*, A, B, C) → R where R is a (relational) concept with learned meaning, while '→' is a basic relation with built-in meaning.

need to directly process the copulas, as all the other relations can be converted to them equivalently.

The knowledge structure of an AERA agent can be seen as a graph of graphs (hypergraph)—sets of linked nodes that hold *explicit patterns*, and the majority of links are latent (implicit), at any point in time, and only those considered potentially relevant at any point in time are made explicit on demand, at run-time, through active comparison and pattern matching, as mentioned above. This is both how concepts are used and formed over time in AERA. Generally speaking, each node (e.g. a CRM or part of it) may contain (small sets of) variables or terms, some with particular values or ranges, but they mostly also reference other nodes, forming a hierarchy. Concepts, in this approach, are thus associated (both loosely and strongly) *subsets* of the full network, where most of the associations/relations are, again, latent, only to be created upon their use for a particular purpose in particular contexts, through said active pattern matching.

Like in NARS, the knowledge in AERA is experience-grounded, and learning, reasoning, planning, and sensing, are always-on continually-running processes. The vast majority of knowledge in an AERA-based system, including causal-relational models (CRM), are abstraction-neutral, and datatype-neutral and defeasible information structures that directly support deduction and abduction. Deduction in AERA is used for (empirical, defeasible) prediction; abduction is used for producing (evidence-based) explanations and (empirical, defeasible) plans.[26] All of these participate in the overt and covert cognitive operations, operating over concept-based knowledge graphs that are generated on demand, as already mentioned. This then collectively defines their (operational) meaning: Overt operations primarily for controlling interaction with the world in which an AERA agent lives, covert operations for steering how its knowledge grows and its cognition develops over time.

The experience-grounded semantics of NARS and AERA can be captured by the following:

1. Statements and information structures are true to a degree, the truth-value indicating the amount of positive and negative evidence collected for a statement or structure via the system's experience.
2. Whether a concept contains a particular instance (example) or property (attribute) is a matter of circumstantial definition (in NARS it depends on the numerical truth-value mentioned previously; in AERA, active deduction and abduction processes).
3. A conceptual relation R can be specified using compound terms (like Cartesian product, as shown in the previous example in NARS on page 28).
4. A compound term and its components are related by the definition of its compound elements, deriving its meaning from these and its active context (relation to other compound information structures at a particular time).

[26] Induction primarily enters into the picture in the creation of new concepts, although AERA can also use it during knowledge compaction (compression). Ongoing work involves the use of induction also in dynamic analogy-making [78].

5. Events experienced by the system can be temporally or spatially related to each other, and these relations also construct compound terms or information structures.
6. The meaning of a term involved in sensorimotor operations is defined partially by its operation, which is procedural by nature. Executable operations have an associated procedural meaning that reveals what they *do;* these operations can be innate or acquired.

According to CCKR, the meaning of a concept changes over time, not only because the system usually doesn't have the time to consider the "full meaning" of every concept when using it, but also because most of the conceptual relations are irrelevant to the tasks the system is working on at any point in time. For instance, a concept C may be linked with 1000 other concepts at a particular moment, those 1000 links (relations) is what C "means" to the system in general at that moment. However, when processing a task (say, answering a question), only 3 out of the 1000 may actually be used, which defines its "meaning for the current purpose," C'. This C' is "constructed on the fly," i.e. composed from the current context, without a specific pointer or name (e.g. a word used to reference it in language), though it is not really a brand new concept, but a sense (usage) of the existing concept C.

As much of the knowledge in the memory will not be relevant at a particular point in time, it is usually the role of resource control/attention mechanism to select the relevant knowledge for processing at any point in time. For instance, for the purpose of comparing a dog's leg to a human leg, features such as "holding up its body," "propelling body forward by pushing," etc. would be selected as points of such comparison, forming a relational graph for the two legs, yielding differences and similarities.

Since the system does not always know how to approach a new situation, rather than rely on existing algorithms, it must use its available knowledge in a case-by-case manner. A system usually cannot afford the resources to exhaustively explore every possible solution to every new problem or situation, due to a general limitation of time and energy, therefore, it must base its actions (mental and physical) on what seems the most relevant relations of its most relevant concepts, at any point in time [84,99], which may necessitate relying on analogies and generalizations.

In NARS, all the knowledge about a particular concept, i.e., all of its links to other concepts, are stored in a probabilistic priority queue. When a concept is used, such as in the process of answering a question, these links are considered one by one to see if it leads to a candidate answer. Each time a link is requested, every existing one has a chance to be chosen, depending on its *quality*, *usefulness* in history, and *relevance* to the current context.

Since these contextual factors may change each time the same concept is accessed and used, the pieces of knowledge in it may be retrieved in a different order, and the amount of total knowledge retrieved for a given task depends on the available time, which also changes from situation to situation. Consequently, the "current meaning" of a concept may be more or less different from its "general meaning." When a concept becomes relatively stable (after having been used many times in similar ways), a smaller number of its relations may usually be involved when it is used. In such a case, these relations could be considered its "essence" or even "definition."

Unlike traditional concept ontologies in AI, many of which draw a fully connected knowledge graph by putting permanent links between shared nodes (and most often created by humans), a knowledge base in AERA creates such node linkage – under a particular context at a particular point in time – via matching *candidate* shared nodes (other nodes with shared properties); if a match is created, it becomes a temporarily-shared node of two temporary links in a temporary sub-graph.[27] For a particular context, some nodes may thus be considered (temporarily) related, while under other circumstances they would not. The key reason for this approach is to more flexibly handle novelty: A difference in how matching, and thus linking, is done enables adaptation of existing knowledge to an agent's present (novel) situation and active goals, which may be different from anything that the agent has encountered in unpredictable ways. Thus, most "graphs" in an AERA knowledge base are not, strictly speaking, "existing" so much as latent and "emergent;" any AERA knowledge base supports numerous graphs for numerous purposes through such dynamic node (and link) matching, on-demand. The meaning is therefore also emergent, depending on the current context and the results of this dynamic graph generation through the appropriate matching processes.

4.3 The Creation of a Concept in CCKR

A major difference between CCKR and the traditional concept-level representation approaches is the assumption that concepts are created from scratch by the system, based on its experience, using its existing knowledge and processes. A direct result of this, and another important difference, is that rather than being predetermined by its designer or trained to converge to a stable structure from the outset, the conceptual network in CCKR is adaptive and constantly changing throughout the system's life time. As the system runs, new concepts and conceptual relations are created constantly and old ones may be removed, to free up resources.

How are *new* concepts created? The typical process of constituting concepts proceeds by assembling (associating) small pieces of information (e.g. data coming from sensors and prior knowledge related to these) to improve the system's performance in various cognitive purposes. For instance, the concept of "drinking from a cup" could be assembled out of existing re-purposed knowledge as "pouring liquid into a mouth," and might include the cup's curvature, how it may be grasped and moved, and how it holds liquid that pours out in the direction it's tilted, as required by the shape of the mouth of the drinker.

[27] Note that these are actually implemented through instantiation and binding of templates; thus, a trace of such events can be stored to support later processing, for instance episodic memory.

In particular, new concepts in NARS come into being in three ways:

1. *Novel external experience.* Initially, the memory of NARS only contains a small number of concepts used in the innate operations. Since NARS is always open to input information expressible in Narsese or perceivable by the sensors, new information constantly enters the system's experience, including novel terms and patterns. If such an item can pass the initial competition and enter the memory, a new concept will be created for it.
2. *Constructions of the system.* NARS has a set of inference rules to organize the experience efficiently. For example, when the system notices a light that is red, it will coin a "red light" concept to combine this two pieces of information into one.
3. *Meaning-drift* of existing concepts. Besides the changing topological structure of the concept network, the numerical features (such as truth-values and priority-values) of the vertices and edges are also adjusted from time to time. In the system's life cycle, the state of the concept network never repeats. Since the meaning of a concept is determined by its relations with other concepts, all these changes will more or less change the meaning of the concepts. When the meaning of a concept has changed significantly, very often it is more natural to be taken as a different (new) concept. This kind of *meaning drifting* is gradual and continuous, so it is hard to decide the exact moment when one concept becomes another. Even so, this phenomenon is inevitable in CCKR, which produce both desired results (e.g., adaptation) and undesired ones (e.g., inconsistency).

As the future is unknown, NARS never knows for sure how much each new concept can contribute to its future activities. Consequently, concepts compete for the limited resources (processing time and storage space). As the system matures, the number of concepts in the system and the number of relations in each concept will be roughly constant, bounded by the system's storage and processing capacity.

In AERA, concepts are not contained in static graphs, as mentioned earlier, but constituted on-demand through a dynamic construction process, in light of the agent's active goals and situation. The rules for assembling the concept graphs use (low-level) pattern matching and reasoning. In this approach, concepts are an *emergent second-order* phenomenon. Introspection on how concepts are constructed – i.e. processes that regulate the second-order concept construction – are third-order processes, corresponding (roughly) to what has been called "cognitive development."

The mechanisms for graph creation in AERA, and thus concept formation, include **contextual association** by *similarity*, *salience*, *urgency*, and *usefulness*. The strength of association between any particular subsets of knowledge, at any point in time, is then a product of these four dimensions. The product may itself be context-dependent, but this puts a burden on the bootstrapping of the learning process (unless the agent has a teacher that can help with the bootstrapping), which depends on an initial seed [84] (i.e. the program that initially gets the learning going).[28]

In both NARS and AERA, this approach makes concepts (a) refinable over time, as experience grows (for instance, how a particular word is used), (b) adapted to the current situation and use, and (c) naturally combined with other relevant concepts, for any purpose, on demand (cf. winged basketballs). This also has the benefit that the cognitive mechanisms used for creating concepts are the very same that use them, avoiding infinite regression of machinery for generating meaning.[29]

In AERA, concept assembly may be initiated from a goal whose end state (e.g., quenching one's thirst) is related to an elements presence (e.g. a cup's presence) in the sensory field: Sensed patterns indicating the shape and size of an object (such as a cup, and the liquid it holds), brings relevant knowledge to bear on the goal (via association); predicted goal achievement enables the first steps of a plan (e.g. reaching for the cup) to be produced and committed to. This exact process is also how brand-new concepts are in fact created: By associating patterns with other patterns. We hypothesize that the flexibility with which a cognitive architecture can do this, and the arbitrariness to which it can be done, determines to a significant extent its capacity for abstract thought.

But if concepts are always created on the fly, how could they reach any stability, and the learner employ them reliably for thinking? How could polymorphic and complex – but useful – concepts, like "water" and "living beings," achieve concreteness and stability in such an approach? The answer lies in the emergent interaction between the on-line/on-demand assembly process, the agent's (dynamic) task-environment, and the fine-grain constituents of the agent's knowledge (low-level representation). The persistence or solidity of a concept is determined by the strength of latent associations between its constituents, which is to an important extent determined by permanent regularities in the learner's world; the association can be temporally local (as, for instance, the concept of a 'winged basketball') or deeply entrenched (e.g., the concept of a human face, parts and all).

In the CCKR approach, concepts are thus not only constructed incrementally and piece-wise by the learning system over time, based on experience, they are also more or less re-constructed every time they are used. As the association between a particular situational factors and particular sets of knowledge becomes stronger, due to repeated experiential evidence of their usefulness (among other

[28] The greater the uncertainty about the task-environment for an agent at "birth," the more general-purpose the initial bootstrapping learning programs must be, at the cost of learning time (the larger the number of options one is open to, more considerations are required before you can know what to pay attention to, and thus the longer it takes to start learning). Thus, the more of the agent's learning machinery that is supposed to adapt to circumstances, i.e. learned, the longer it takes to stabilize.

[29] While it is tempting to go into further detail on this particular topic, it involves a number of runtime and implementation issues that are unfortunately too involved to be given a treatment here.

things), the more likely it is that mostly the *same* elemental knowledge will be used for creating *similar* graphs in *similar* situations. Situations with an overlap – especially situations that differ wildly, yet associate with a small subset of the same knowledge – will help highlight that particular subset of knowledge as being relevant across otherwise different situations. This has the desired side-effect of *concept solidification*. Furthermore, the knowledge thus created will become increasingly useful for particular purposes in those situations over time, because it is constantly being improved through experience, and this in turn incrementally increases its conceptual contextualized solidification.[30]

Learning proceeds, in the CCKR framework, by explicitly modeling[31] selected subsets of ongoing experience, resulting in information structures that can (directly or indirectly) inform a learner's future behavior, both overt or covert. The adequacy of any information structure thus created, in light of a particular goal, c.f. "knowledge X can be used for achieving goal Y," must take into consideration the context in which it is to be applied, i.e. "in situation Z".[32] This means that any learning controller that achieves autonomy through the use of explicit information structures in a particular environment – that is, freedom from going back to the lab for re-programming when facing novelty – must also have adequate means for their creation, modification, continuous improvement, and for evaluating their usefulness for various purposes in various contexts. This last part, in our approach, should be achieved through a recursive application of *the same concept-centered mechanisms* being thus evaluated, to avoid infinite growth of the information storage.[33]

---

[30] We hypothesize that this is in fact where the power of language to shape cognition comes from: It is, among other things, a "concept solidifier" mechanism—making certain concepts more useful than others, as evidenced through communicative experience. It may also be the source of some confusion with a hypothesized "language of thought" [21], which in our view is most aptly thought of not as a language but as a reasoning control system.

[31] The term 'model' is used here in the most general sense, as "the strategically chosen features, functions, and relationships in the learner's (overt and covert) experience, purposefully reformulated as information structures" that are useful for achieving an agent's implicit and explicit goals.

[32] This matches the general definition of a *model*, that is, a model with a referent (whether we assume the referent being something out in the physical world or simply the experience of the learner). For instance, a sowing thread may not be useful as a model for answering questions about the strength of a tree trunk, but it might work for answering whether the tree is long enough to bridge the banks of a river).

[33] One critique of this might be that we are replacing infinite growth with infinite recursion, which would in fact be correct. Of these two evils we prefer to take on the latter challenge rather than the former, because (a) it seems to us that humans can apply their concept-creation abilities to virtually anything, a seemingly important feature of human cognition, and, (b) while humans may be able to create apparently infinite recursions of the kind "I know that you know that I know that you know that I know...," they seem to be quite capable of avoiding it becoming an impediment to organized thought.

It may be useful here to sidestep ever so quickly into human cognition with respect to language. Explaining the role of language in concept use and thought is still a largely unsolved question of human intelligence. In CCKR the mechanisms of language creation and use are directly based on the concept creation and use we have described here. In this view, a word is a particular pattern[34] (whether written or spoken) that can be associated with a particular knowledge subset, through the mechanisms outlined above. The result is that any useful association between that pattern and a set of information structures will solidify over time, upon repeated use and experiential evidence of practicality. Importantly, this solidification is *local* in that due to the compositionality of concepts, only elements hypothesized to be *strictly relevant* to the goals and situation at hand are consolidated, as are subsets of the concepts thus created (i.e. concepts are not "blobs" but rather, "Lego sets"). Context plays an important role here: In AERA the knowledge deemed relevant upon reading or hearing a word will depend on the current situation (determined by input from the sensors) and the currently active goals, controlled by the system's resource management mechanisms. Due to the way its knowledge is encoded and its acquisition proceeds, to teach AERA to use words – and language – is to use them in context. Once learned, AERA can use not only appropriate words but also appropriate syntactic patterns to describe or talk about any of its known concepts, treating the words, e.g., "table," as a pattern whose relationship to models of the concept 'table' is described by *models of the **use** of the word* "table." As the same symbol (word pattern) is subject to the same knowledge management mechanisms as any other knowledge, procedural or otherwise, a word like "car" will naturally result in slightly different graph construction depending on whether the context is the 19th century or the 21st century. With training, words become useful for thinking, because of their rich associations to other knowledge—and they recall contextually relevant knowledge, just as the sight of a cup will recall concepts related to holding cups, drinking from cups, etc., if your goal is to find a cup, and only if the cup is green when you're looking for "green things."

4.4 CCKR: Properties and Implications

A concept-based cognitive general learner in a complex world creates (and improves) its knowledge piecemeal, from small atomic ("peewee") elements, over time. Acquired knowledge is thus made from components (sub-parts) whose meaning is defined by, and emerges from, their use in bridging between sensation and action, prediction and plan, explanation and problem solving, abstraction and generalization, and so on. In the process of their application for these purposes, the concepts and their relationship with other concepts is reinforced (and re-instated) through the existent relations between their constituent components. Seen this way, a concept is a piece of segmented and compressed multi-faceted model of experience.

[34] To be more specific, a word rightfully forms a *family* of patterns, due to the tolerance for a human language user to its variations in practical scenarios.

A concept in CCKR is thus a collection of information structures that are internally related, and describe multiple aspects of a "semi-unified whole" of closely-related things that are relevant to a being's cognition, including its active goals and proximal immediate surroundings and actions. The concept of a 'table' includes (latent) models of its physical features, what it's for, how it's constructed, etc. It includes all the invariant information that can be abstracted from a set of (experienced and hypothesized) instances of tables (i.e. where the label "table" is appropriate, based in part on social convention). Edge cases test these conventions and the information structures that *produce* the concept, so that's why people typically have to *think* more when asked whether something that's made from unusual materials or is missing some standard features is a an instance of that concept. For instance, is a waist-high flat rock in the forest a table? That depends – in part – on the role of the rock in your activities: the very act of deciding whether something "is a table" (and thus also whether to call it, use it, or treat it as a table) is handled by similar mechanisms that enable the creation of new concepts.

The knowledge contained in a CCKR database can be inspectable and manipulable at a fine-grain "atomic" level[35] because the relations between parts of the knowledge are constructive—i.e. they have a decomposable structure whose parts can be inspected – reflected on – by the system itself. In a given conceptual hierarchy, at the lowest level of any concept are its atomic peewee elements – these cannot be dissected because they don't contain anything smaller-grain. Some atomic elements may be expanded by further learning; for instance, even if the wings of an actual airplane are no more detailed in our mind than those of a paper airplane, we may extend it further when learning that there are flaps at the back edge that can move and internal tanks for holding fuel. Other atomic elements may terminate in some fundamental measurement that can be provided by a controller's transducers, and thus usually pertain to an agent's I/O devices. These are key in grounding all learned concepts, as they play a fundamental role in bootstrapping the agent's learning process itself.

A concept, in this conceptualization, can be a compound information structure composed of other concepts, patterns or partial concepts – that may be manipulated as a set, all at once, while also allowing dissection and analysis, according to needs (decided by an agent's current situation and goals). Concepts are thus clusters of introspectable knowledge that contains overlapping graphs that can be dissected down to the smallest atomic pieces of knowledge.

For a large knowledge base, containing thousands or millions of low-level concepts (e.g., how the apparent shape of objects depends on viewing angle, to take an example), the combinatorics and potential to form sub-graphs will

[35] An analogy to physical atoms – not their Platonic indivisible original version – is somewhat useful here: Physical atoms constitute not only something of a coherent whole, they also inherently encapsulate the rules about which they "play" with their surroundings, which atoms they may pair with, and so on. In this sense, like atoms, models in AERA and terms in NARS are made of parts that determine their behavior in the context of other things.

approach infinity. To be flexible – i.e. to be useful in as many situations for as many purposes as possible, supporting a large variety of potential knowledge graphs – elements of such graphs would need to be re-used across various situations (e.g., the 'supporting' role of a human leg is similar to the 'supporting' role of a table's leg—and probably partly responsible for them both being called 'legs.'). For a large knowledge base it is of course prohibitive to pre-compute every possible information structure that might be needed, and in a complex dynamic world, knowing beforehand everything might be needed in the future is also impossible. These are three reasons why our approach relies on dynamic contextual concept (re-)constitution: Dynamic (re-)creation means that the relevant building blocks can be used for thinking, in light of *any* relevant purpose and situation, making knowledge use very contextualizable. This is also why reflection is called for: A cognitive system must be able to inspect its own operation, to the extent necessary to allow for part-wise manipulation, modification, comparison, and evaluation of any subsets of its knowledge.

The latent concepts in a large knowledge base will be highly context-dependent, putting into question the idea that when a concept is used it is the "same" concept as when it was used last time. This is a feature, not a bug: It means that the relevant parts of a concept are summoned for particular intended purposes. Human concepts exhibit this in well-studied edge cases, such as whether a broken flower vase is still a vase (the answer depends in part on what you plan to do with it—are you testing the power of a new kind of glue or do you have some flowers that need to be put in water?). Many examples of fluid and creative examples of usage can be found in [35].

5 CCKR: Summary of Key Features

CCKR concepts bring a computational "economy of thought" [46] – an Occam's Razor – that allows a learner to use bounded resources more efficiently. This involves (1) simplification, in the form of summaries of complex experiences, (2) identification of similarities (by grouping), and (3) steering of attention (deciding relevance for efficient steering of resource usage).

A concept-based representation is useful for handling unbounded objects and events, by being domain-free and scale-free, meaning that the ability of compositional concepts to represent information isn't limited to particular domains and they aren't bound to a particular level of spatiotemporal description.

CCKR information structures contain the necessary representational features to be *autonomously formed and managed* by the learner, *continuously and consistently* over its whole lifetime, on demand. CCKR knowledge is explicit and new concepts can be constructed from existing concepts using clearly defined operators and explicit relations; existing concepts can be split and re-combined, modified, and changed in part or in whole. CCKR information structures naturally represent hierarchies as well as various types of compound concept, even target other CCKR structures in support of reflection. As a result, CCKR representation supports naturally multiple types of reasoning (deduction, abduction, induction, and analogy), and some of them are often considered as rules of learning, too. This means that a CCKR learner can explicitly and systematically compare and contrast its own experience, reason about its own knowledge and cognitive processes, and produce arguments for its actions. This is a fundamental method by which life-long continuous (cumulative) learning can be realised [85].

In principle, the future is different from the past, and every situation is novel. Adaptation means treating anything novel in relation to the familiar, through decomposition, recognizing their partial similarity and ignoring their differences. A concept in our view provides the mechanisms to achieve Piaget's idea of "assimilation" [66] and Hofstadter's "seeing as" [38], allowing for natural decomposition of information gathered over time through experience.

The knowledge a CCKR system obtains is normally from the processing of multiple tasks, so its meaning, usefulness, and significance to the system is not limited to a single task, as that obtained in an end-to-end black-box learning process that aims at the approximation or optimization of a specific function [84,107].

Since CCKR represents knowledge as concepts and their relations, a learning system using it will produce knowledge that is easier to align with human concepts than most other approaches to knowledge representation, and therefore easier for humans to understand; it also makes it simpler for a learning system to combine its own experience-based knowledge with given human knowledge.

6 Comparison and Discussion

In the following section, CCKR is compared with other approaches of knowledge representation, especially the symbolic and connectionist schools.

6.1 Theoretical Underpinnings and Methodology

In light of the bigger historical picture, research on representation for intelligence is still in the early stages, which means that a large number of important background assumptions and requirements have yet to be addressed by mainstream research, and many have in fact yet to be brought into the foreground [88]. This includes the theoretical underpinnings of the methodologies. Indications that we are in the early stages of research can be seen in the relatively high number of cognitive features that are absent from most current AI systems—it is for instance still not uncommon to see cognitive architectures that are incapable of learning, and very few of them handle time as a first-class citizen. As there exists no accepted broad theory of intelligence, this is perhaps not surprising. Instead of working towards such a theory, however, which requires elucidating all main background assumptions, the research community has addressed its subject matter largely by fragmenting along topical-methodological lines – neural, symbolic, etc. – and researchers in each camp continue to speak to a (relatively small) community with a limited set of shared background assumptions and terminology, which in turn then remain generally obscure in the larger community. This is the current situation, and it could easily go on unchanged for decades due to the complexity and intricacy of the topic.

To make faster progress, a concerted effort to bridge between sub-fields, with common terminology and definitions, must be undertaken. A simple but important step includes stating clearly in research papers the background assumptions and methodological approach followed. It may be argued that connectionist approaches to AI are not based on any particular theory of intelligence or cognition at all but rather, on an extensive analogy—that "brains are the machinery" of intelligence, so by copying (or rather, taking inspiration from) a subset of its features we can re-create intelligence in a machine. While this makes perfect sense for studying the fine details of brain operation, such as neurons and neural clusters, it is not very useful for understanding the higher levels of organization necessary for cognition—and certainly not the level of human knowledge and concepts. When studying human cognition, *minds* and *brains* are certainly two aspects of a particular incarnation of the same phenomenon, but their relationship is not such that one can be naturally and easily derived from the other. A mapping between these two levels ought to be possible, and will undoubtedly be achieved eventually, but it is far from straightforward. Without a theory of intelligence, it is quite a conundrum which of the thousands – or millions – of biological feature candidates in a working brain should go into a successful implementation of an artificial mind.

It may further be argued that the symbolic approach in AI rests on essentially the same exact theoretical basis as that commonly accepted for computer science. This might not be surprising, considering that historically the origins of the basic ideas on which both are based trace back to approximately the same period in history, when practical electronics, information science, and software development were forming, shortly before and around the middle of last century. "The representations used in computation are symbols and the mind is a computation—the human mind must thus also operate on symbols." No wonder that the founding fathers of AI should think that computers would be able to demonstrate human-level cognition within a decade! But now we know they didn't—not by a long shot.

If current theories of computation suffice to describe computation fully, both practically and theoretically, a theory of cognition cannot be the same as that of computation, whether it be the particulars of the information architecture upon which cognition rests – like is commonly thought – or something else, because they are fundamentally different; the principles allowing human thought to surpass contemporary theories of computation in fundamental ways are still missing. But what could the principles of cognition be, beyond those which describe computers?

Our CCKR approach rests on a theoretical basis for (artificial) intelligence that has some overarching principles to offer that go beyond current theories. It rests on a working definition of intelligence that assumes intelligence fundamentally involves adaptation under insufficient knowledge and resources [83,104]. It follows from this stance that intelligence operates under *relative rationality* [101], and that intelligent agents acquire their own knowledge through *active construction* of information structures [82]. While the latter point certainly echoes the work of some older psychological theories and schools of thought, our work takes several important steps that, unlike this early work, allows the specification of building instructions for machines that capture these essential properties.

In narrow AI, as in traditional software development, the meaning of anything that an artificial system does can only be interpreted from the point of view of the human creator of that system. This is one key reason why people are often reluctant to attribute understanding or intent to, for instance, AI systems like self-driving cars; in the words of the proverbial man-on-the-street, a self-driving car can't be made responsible for its actions because it doesn't really "know what it's doing." Put differently, its actions don't have any *meaning to itself*—a self-driving car doesn't even have any *handle* on meaning: The context of its actions are not part of its operation. It is a tool, plain and simple, unavoidably resulting from the allonomic engineering approach used in its construction – as with all traditional software development [82].

In contrast to traditional software development methodologies, our approach requires a learning agent to create its own meaning. It does so in the way in that it creates and uses information: The knowledge of an agent with *real* intelligence is not an accurate and objective description of the world, but rather, a particular *encoding of the agent's experience*, with *concepts* as reusable units. The knowledge of such a learning agent is always a *construct*, and its learning thus *constructivist*. It also means that while knowledge is constructed, it is in fact *hypothesized*, because the learner can never know what is "truly out there." This has the important and unavoidable conclusion that knowledge is *never certain* and *all reasoning* that such an agent does, based on this knowledge, is *defeasible* [69,70] and *non-axiomatic* [93].

Learning, in this view, consists of the construction of interconnected (and partial) models of a learner's accumulated experience, about its own actions, including its own cognitive events (thinking about thinking), and those of the world, in a "language of thought" that supports recursion (i.e., one that is compatible with other knowledge representation and mechanisms). If there is regularity in how the agent's sensors respond to physical forces in the world, the models will reflect features of the agent's environment, as sampled through experience, in a compressed form (which depends in part on the agent's cognitive apparatus; the other part it depends on is the form of the agent's sensors and history of interaction with the environment). Over time this knowledge becomes increasingly useful for meeting the agent's numerous active goal(s). The above considerations form a foundation that separates, in our view, intelligence found in nature from current narrow AI, whether symbolic, sub-symbolic, or something else. They result in a set of implications that, again in our view, makes it clear that current approaches to AI are unlikely to lead to AGI or "real AI." Attempting to put these in a comprehensive list is thus our next task.

6.2 CCKR Concepts vs. Symbols

A predictable misunderstanding of what we have presented is to take CCKR as yet another symbolic approach, since a concept may look a lot like a symbol, which also can serve as an identifier of information structures within a system.

We insist that the concepts in CCKR are not symbols, mainly because of the fundamental difference in how they work, and what a concept and a symbol each *mean*. In the view of CCKR, symbols do exist, but not in the mind: Instead, they are the physically manifested representation of a concept, thought, or relationship – or a sequence thereof – such as a word (composed of letters), a "no-smoking" sign (composed of symbols), a particular temporal sequence of sound waves (speech), etc. Such physical symbols are used for communication to allow the creation of a shared set of concepts, relations, etc., through sensory apparati—physical symbols are manifestations of thought based on concepts.[36]

Further, if 'symbolic' were to be interpreted as equivalent to "using identifiers," the notion would apply to all systems implemented in digital computers, including the connectionist models, as the nodes and links are addressed using internal identifiers or names. So this is clearly inadequate, and a more thorough dissection is called for.

In their "physical symbol system hypothesis," Newell and Simon [56] stated that a symbol "designates" an external object or a process, and the symbol can be "interpreted" according to this mapping. Given this relation, the system's manipulation of the symbol will make it behave according to the object outside the system. This usage of "symbol" follows the referential view of semantics [63], in accordance with traditional *model-theoretic semantics* [6], a well-established theory. This semantics has played a central role in mathematics for specifying the meaning of mathematical concepts, but its application in AI has significant shortcomings. As a result, it has received its share of criticism such as Searle's "Chinese Room" thought experiment [76], which rightfully highlights a major blind spot of all symbolic (GOFAI) AI approaches in that a system using symbols according to model-theoretic semantics has no access to the interpretation of the symbols in it, because according to that theory, the meaning of the symbols is determined by an interpretation that is independent of the system's experience, or the way in which the symbols are processed by the system. The issue is often expressed as the "symbol grounding problem," i.e., the question *"how can the semantic interpretation of a formal symbol system be made intrinsic to the system?"* [29, p. 335].

CCKR is not a solution to this problem but a rejection of its presumptions – that is, of the very foundations of model-theoretic semantics – which define the meaning of a symbol by its denotation outside the system. In CCKR, the meaning of a concept is determined by its role in the system's experience, as revealed by its relations with other concepts [95] and its use for overt and covert action [84]. For this to be possible, the creation, manipulation, and management

[36] The creation, manipulation, comparison, etc. of concepts is, however, "symbolic" in the sense of "using identifiers;" see following paragraph.

of concepts must be accessible to the system and automated in the system's operation, i.e., their meaning must be determined by the system itself at runtime. In this view there is no designation or interpretation from an observer. As long as a concept appears in (or can be related to) the system's experience, it has some (latent[37]) meaning to the system. Here the "experience" can be anything represented as internal events—linguistic, sensorimotor, or anything else that the system experiences, including covert operations, and the relation can be direct or indirect, i.e., via zero or multiple abstractions from the raw input and internal operations.

Specifying the meaning of a concept or a word by its relation to other concepts/words is not a new idea, as proposed in *conceptual role semantics* [28] and the *language-game* theory [111]. In computer science and AI, the approach has taken various forms, from semantic network [112] to knowledge graph [19]. These "relation-based" approaches have mockingly been termed "Dictionary-Go-Round" and criticized as defining a meaningless symbol using other meaningless symbols. This is not the case in CCKR, because here the conceptual relations are not definitions, but rather *experienced* by the system itself—we could perhaps call it "contextuo-temporally grounded" because in this view any experience is couched in the context of internal operations and similar experiences, temporally (and spatially) demarcated.

CCKR requires the concepts to be (eventually) related by the built-in relations that are directly processed by the system, and thus have an in-born operational semantics (seed-based knowledge in AERA; copulas in NARS). The built-in relations do not require any interpretation to become meaningful, and neither do they allow multiple interpretations, as they are used according to the meta-level knowledge generation processes innate to the system [84,97]. These (innate) concepts have fixed meaning that is not influenced by the system's experience, and they also contribute in determining the meaning of the other (acquired) concepts using the system's experience.

Another common criticism to the relation-based approaches is that a concept can be related to too many other concepts for such a representation to be practical. This issue is resolved in CCKR by its acknowledgment of resource restrictions and dependence on an attention-allocation mechanism, which limits selected relations at the point that each concept is used to the relevant ones for an immediate purpose (temporally defined by the situation and active goals). This attention mechanism decides the current meaning among the general meaning of a concept, and therefore also explains the context-sensitivity of concept usage. This will be controlled by various cognitive operations, many being some form of reasoning, will determine the particulars of how this proceeds.

An important property of CCKR is its dynamic nature. As the conceptual network comes from the system's own experience, it will continue to change

[37] The meaning is latent because one or more such meanings may not be realized for the purposes of a particular situation at a particular time—for this reason it might be better to say that 'a particular concept may have many latent meanings, and specific meaning(s) depending on its use in a particular situation.'

as long as the system is running. In contrast, traditional symbolic AI systems typically assume that every symbol has a fixed meaning that is given by its (fixed) interpretation, being independent of the system's knowledge, history, active goals, and current status.

The claim here is not that model-theoretic semantics is *completely* incompatible with CCKR,[38] but rather, that it should not be applied to the acquired and empirical concepts [82,98]. The most important similarity of CCKR and symbolic AI is at their level of abstraction, that is, trying to build a *mind*, rather than a *brain*. However, they make very different assumptions about how a mind is related to its environment and what knowledge is: CCKR models the system's interaction with the world, rather than the world "as it really is."

6.3 CCKR Conceptual Networks vs. Artificial Neural Networks

At a high level of abstraction, CCKR and artificial neural networks (ANNs) share the common features of being networks (graphs) consisting of interconnected units, where the knowledge is distributed in the (semi-explicit) edges, which have "weight values" that can be learned. When the system runs, there is a form of "activation spreading" along these edges. This is mostly where the similarities end, however; CCKR is otherwise very different from ANNs and other connectionist models in many important aspects.

CCKR and ANNs correspond to very different levels of description of the brain, that is, the (psychological) *conceptual* level and the (biological) *neural* level, respectively. A widely-used argument in favor of the ANN approach is that the brain is a neural network, so this is proof that a network-based approach is right for creating intelligence (cf. [77]). The statement is severely misleading, however, in that even though the brain could be fully described as a neural network (at one level of abstraction), this network would still be very different from current ANNs, and it does not exhaust all aspects of the brain functions – a brain is not *merely* a neural network. Furthermore, the validity of this description level does not exclude others. In particular, human introspective experience of our own thinking process is solely at the concept level (we say "I get this idea," not "I have this neural cluster activated"). This is one of the fundamental reasons why the ANN approach to AI runs into trouble when a knowledge-level explanation of their processes is demanded, because we do not understand neural activities (in either their biological forms or their mathematical imitations) at the level where our own learning, reasoning, and thinking are usually explained.

It can be argued that these two levels are related but that the neural level is in a sense "more fundamental." Even if this were the case (which we find a tenuous claim), it does not mean that conceptual-level descriptions have to be reduced to the neural level to be meaningful or useful, or even correct. On the contrary, conceptual level descriptions have the advantage of being independent

[38] There can still be symbols whose meaning is decided by an interpretation, as those in seed-models in AERA, and those used in the axiomatic subsystems in NARS.

to the biological or evolutionary aspects of the human brain, which are not necessarily needed for intelligence in general. Few would argue that the operations of a word processing program should be reduced to the transistor level to be understandable or meaningful. Intelligence is an information process, and any theory of it must address this level at a minimum [55]; bridging to lower levels – explaining how they can be implemented – is an added bonus. A historical reason for why many people prefer neural networks as the theoretical underpinnings of their work is because they see it as a promising alternative to the symbolic representation, which historically has been rigid, fragile, and dependent on human construction [34,37,80], and its resemblance to human brain is secondary. Based exclusively on the symbolic AI work of the 60 s and 70 s, such a view is understandable, but is an unfortunate results of incorrect induction from limited historic events.

Some researchers may argue that there are also 'concepts' in an ANN. Though in a sense the neurons at the input and output layers do share some properties with the concepts in CCKR as "identifiable information structure," they lack the rich semantic, dynamic, and constructive features stressed in CCKR.

Bringing many of the good features of both symbolic AI and ANNs [96], our CCKR approach offers the following desired features:

1. Fluid, adaptive, and flexible ("stretchable") concepts: In CCKR, the boundaries of concepts are not sharp or binary, that is, whether a particular instance of a phenomenon belong to a particular concept is a matter of degree, adjusted by the system itself according to history, purpose, and context.
2. Tolerance to various types of uncertainty: CCKR allows uncertainty in knowledge, so local or global inaccuracies, incompleteness, and inconsistencies will not crush the system.
3. Parallel processing: CCKR allows multiple tasks to be processed concurrently, either in a time-sharing manner or in multiple processors, because it does not require global consistency among the system's beliefs or desires (though the system always tries to maintain these consistencies as much as it can).
4. Self-organization: CCKR supports the construction of new concepts in multi-level abstractions, as feature learning in deep neural networks (DNNs). Unlike DNNs, however, the concepts are made from parts that themselves are individually addressable, inspectable, and expandable, enabling self-organizing compositional concept hierarchies.

These have all been demonstrated in some way in our work on NARS and AERA. Equally importantly, these benefits do not come at a cost but rather, CCKR avoids some major weaknesses exhibited by prior approaches, including ANNs [96]:

- Compositionality: CCKR knowledge is explicit and new concepts can be constructed from existing concepts using clearly defined operators and explicit relations; existing concepts can be split and re-combined, modified, and changed in part or in whole. CCKR information structures naturally represent hierarchies as well as various types of compound concept, even target other CCKR structures in support of reflection.

- Autonomy: CCKR information structures contain the necessary representational features to be *autonomously formed and managed* by the learner, *continuously and consistently* over its whole lifetime, on demand.
- Generality: The knowledge a CCKR system obtains is normally from the processing of multiple tasks, so its meaning, usefulness, and significance to the system is not limited to a single task, as that obtained in an end-to-end black-box learning process that aims at the approximation or optimization of a specific function [84, 107].
- Reasoning: CCKR representation is in large part explicit, and supports naturally many forms of reasoning, including deduction, abduction, induction, and analogy, on any knowledge subset. This means that a CCKR learner can explicitly and systematically compare and contrast its own experience, reason about its own knowledge and cognitive processes, and produce arguments for its actions. This is a fundamental method by which life-long continuous (cumulative) learning can be realised [85].
- Explainability: Since CCKR represents knowledge as concepts and their relations, a learning system using it will produce knowledge that is easier to align with human concepts than most other approaches to knowledge representation, and therefore easier for humans to understand; it also makes it simpler for a learning system to combine its own experience-based knowledge with given human knowledge.

6.4 Unification vs. Integration

CCKR is based on the hypothesis that intelligence in general can be fully explained and reproduced at the conceptual level, largely independently of its substrate and underlying lower levels, being it biological neural networks, electrical circuits, or something else. CCKR does not aim at describing human knowledge representation specifically, or in detail, but to abstract its fundamental principles to a level where they are equally applicable to a wide range of computer and natural systems.[39]

While sharing features with both the symbolic and the connectionist approaches, CCKR is neither a hybrid nor an integration of them, but rather a new and broader proposal, resting on a reasoned foundation, that subsumes their best features and overcomes their individual limitations. The resulting CCKR theory does not contain "a symbolic part" and "a connectionist part" but rather, breaks the walls between these two paradigms and gets at the more fundamental principles that their commonalities rest on, achieving a larger, more coherent picture.

One fundamental reason for the failure of past attempts at a proper theory of concepts is that many of the strong theoretical principles on which prior

[39] Of course, eventually it needs to be explained how such a representation is implemented in the human brain, and how it differs from other animals, but that is neither a topic we consider to be fruitful as a primary methodological principle for AI research in its current state nor one to be addressed in this article.

attempts have rested are mostly non-existent here, or only present in a weaker form. Without a good (or good enough) theory, attempts at unification ultimately must make do with mere integration instead, where an improper combination inevitably results from incompatible assumptions about the operation of the whole. Yet it is precisely *that which is lacking* that would allow a comprehensive unification: NARS and AERA do not have separate "symbolic modules" and "connectionist modules," nor do they directly use existing logic or ANNs. Instead, they can be considered as the traditional logical or rule-based systems completely rebuilt according to many subsymbolic/connectionist ideas [37,80,96,107].

We agree with many other researchers that a hybrid or integrated approach may be more suitable (in terms of simplicity and efficiency, for instance) for certain specific problems, though it will be difficult, if not impossible, to build a general-purpose intelligent system (AGI) using such an approach. It is true that both paradigms can find supporting evidence in human cognition, as the same cognitive process can be described at either the conceptual level (as in psychology) or the neural level (as in neuroscience), though neither can claim to have captured all the information about the process.

Since connectionist and symbolic approaches correspond to different levels of abstraction when referring to human cognition, each uses a unique vocabulary to explain target phenomena in a somewhat incompatible way. There surely is a correspondence between the neural-level and conceptual-level descriptions of the same process, but it is not a one-to-one mapping; in general, there is no direct mapping between neurons and concepts, nor is there a separate "neural part" and "symbolic part" in the brain. CCKR is 'unified' also because it uses 'concept' in a broad sense to cover the territories which are traditionally referred to as "subsymbolic" or "perceptual." As described previously, since according to CCKR a concept is an internal entity of varying size that can be explicitly identified as a unit, there is no reason to exclude mental image, goal, operation, etc. from forming a concept.

CCKR takes *concept* as a central unit of representation, but does not treat its *meaning* as a constant determined by designation, definition, or third-party interpretation, but rather, being contained in the multifaceted properties learned by the system from experience. This position makes it fundamentally different from other approaches in artificial intelligence and machine learning, though is closer to many research on concepts in cognitive science [35,43].

In our approach, the meaning of a concept is determined by the system's experience and context, and thus CCKR is also compatible with the view that cognition is *embodied* [2], because a system's "body" is by definition demarcated by sensorimotor operations, which partially constrain the system's experience. When a learning system is implemented with different sensors and actuators, the system's concepts and beliefs will be different, even though the mechanisms responsible for its intelligence (concept-centered representation, unified inference capability, resource allocation procedures, etc.) may remain the same. The changes in body and experience have an impact on the system's problem-solving skills, but not its intelligence, which is defined at the meta-level of the skills, and is (largely) body-independent [100]. Thus, the requirement of embodiment should neither be understood as a requirement for the meaning of concepts to be reduced to the sensorimotor level, nor the requirement that the system's body (and experience) be close to that of a human [4].

By extension, CCKR also covers *situated* and *context-sensitive* cognition [3], because its memory is dynamically structured to favor the concepts and relations that are directly related to the current situation, or the need raised by it, as explained in [97]. However, it does not mean that the system is exclusively reactive, with no long-term motivations and drives; quite the contrary: the CCKR approach fully supports reasoning, widely construed, including prediction, explanation, causal reasoning, and analogy making (though in this article they cannot be described). Here the *context* of a concept is provided by other related concepts and their current role in light of active goals and situations, rather than by explicit labels, as in the traditional symbolic AI approaches [7,49].

7 Conclusions

Our proposed concept-centered knowledge representation (CCKR) is directly based on our understanding of intelligence, cognition, and mind. While resting on – and echoing – many prior ideas in the field of AI and cognitive science, we consider our formulation new in that it is more specific and offers, and contrasts here, two implementations (NARS and AERA) that have already demonstrated to go beyond other methods in machine learning and AI in important ways. While both are certainly in their early stages, we consider them sufficiently mature to compare to prior attempts at unified theories of cognition. This paper focuses on the information content and form of our proposed concept-centered knowledge representation demonstrated in these systems, but our work extends well beyond these, sufficiently so to allow functionally operational software demonstrations of its key principles.

CCKR does not assume the existence of an external world that consists of clearly separated objects and events waiting for the system to discover, recognize, or manipulate them. On the other hand, this position does not assume an arbitrary or random environment either. We assume that an environment exists independently of the system and is complex enough that its future cannot accurately be predicted in full detail, while containing sufficient regularity for learning through experience, since otherwise no adaptation is possible.

CCKR proposes that in any general-purpose intelligent system, knowledge must be represented as a dynamic conceptual network with particular properties and related operations. This network is neither a "neural" network, as each concept has an internal identifier function that allows the network to be used to construct more complicated concepts, to be constructed, re-constructed and de-constructed, and is therefore not restricted to a specific task or function. Nor is the network a "symbolic" network as classically construed, since the concepts do not refer to external objects and events; the network's structure, as well as the concepts and their relations, constantly change as the system interacts with the environment while accomplishing its tasks through continual reasoning and learning.

CCKR is presented as a set of principles that allow different formalization and implementation, rather than a single model. This is the same case for the symbolic and connectionist approaches, as both of them include various concrete designs, while still having enough features to be distinguished from each other. For this reason, we have not forced CCKR into a single formal model, though it still shows enough differences from the other two.

A concept in our approach is neither a symbol, nor does it necessarily represent an external object or event, but rather, an ingredient, pattern, and model of the system's *partial* experience, that is, a record of its interaction with the environment and a record of its own use of its knowledge. While some basic concepts can be innate or given, most concepts are constructed by the system itself, and all concepts are modifiable by the system according to history and context, and with different degrees of stability and clarity.

The benefits of a concept-centered knowledge representation include:

- Treating novelty through decomposition and partial similarity to what is already known, ignoring and/or examining the differences. This ability is fundamental to all intelligence because the future is always different from the past and every situation is novel.
- More efficient use of resources in representing and learning from experience, bring a 'computational economy of thought:' Concepts allow for (1) simplification, in the form of summaries of detailed, complex experiences, (2) identification of similarities (by grouping), and (3) steering of attention (using estimates of relevance to achieve more efficient steering of resource usage).
- Freedom from predetermined levels of spatiotemporal descriptions: A concept-based representation is domain-free and scale-free, meaning that the compositional concepts can represent information that isn't limited to particular domains or bound to specific level of spatiotemporal description, allowing them to handle unbounded objects and events.

References

1. Anderson, J.R., Lebiere, C.: The Atomic Components of Thought. Lawrence Erlbaum Associates, Mahwah (1998)
2. Anderson, M.L.: Embodied cognition: a field guide. Artif. Intell. **149**(1), 91–130 (2003)
3. Barsalou, L.W.: The instability of graded structure: implications for the nature of concepts. In: Neisser, U. (ed.) Concepts and Conceptual Development: Ecological and intellectual factors in categorization, pp. 101–140. Cambridge University Press, Cambridge (1987)
4. Barsalou, L.W.: Perceptual symbol systems. Behav. Brain Sci. **22**, 577–609 (1999)
5. Barsalou, L.W.: Challenges & opportunities for grounding cognition. J. Cogn. **3**, 1–24 (2020)

6. Barwise, J., Etchemendy, J.: Model-theoretic semantics. In: Posner, M.I. (ed.) Foundations of Cognitive Science, pp. 207–243. MIT Press, Cambridge (1989)
7. Barwise, J., Perry, J.: Situations and Attitudes. MIT Press, Cambridge (1983)
8. Beer, R.: Dynamical approaches to cognitive science. Trends Cogn. Sci. **4**, 91–99 (2000)
9. Besold, T.R., et al.: Neural-symbolic learning and reasoning: a survey and interpretation. arXiv abs/1711.03902 (2017)
10. Birnbaum, L.: Rigor mortis: a response to Nilsson's "Logic and artificial intelligence". Artif. Intell. **47**, 57–77 (1991)
11. Brooks, R.A.: Intelligence without reason. In: Proceedings of the 12th International Joint Conference on Artificial Intelligence, pp. 569–595 (1991)
12. Brooks, R.A.: Intelligence without representation. Artif. Intell. **47**, 139–159 (1991)
13. Brooks, R.A., et al.: Alternative essences of intelligence. In: Proceedings of the Fifteenth AAAI/IAAI Conference, pp. 961–968 (1998)
14. Chomsky, N.: Syntactic Structures. Mouton, The Hague (1957)
15. Conant, R.C., Ashby, W.R.: Every good regulator of a system must be a model of that system. Int. J. Syst. Sci. **1**(2), 89–97 (1970)
16. Dinsmore, J. (ed.): The Symbolic and Connectionist Paradigms: Closing the Gap. Lawrence Erlbaum Associates, Inc. (1992)
17. Dong, H., Mao, J., Lin, T., Wang, C., Li, L., Zhou, D.: Neural logic machines. In: Proceedings of the Seventh International Conference on Learning Representations (ICLR 2019) (2019). https://openreview.net/pdf?id=B1xY-hRctX
18. Dreyfus, H.L.: What Computers Can't Do, Revised Harper and Row, New York (1979)
19. Ehrlinger, L., Wöß, W.: Towards a definition of knowledge graphs. In: Joint Proceedings of the Posters and Demos Track of 12th International Conference on Semantic Systems - SEMANTiCS2016 and 1st International Workshop on Semantic Change & Evolving Semantics, pp. 13–16 (2016)
20. Ensmenger, N.: Is chess the drosophila of artificial intelligence? A social history of an algorithm. Soc. Stud. Sci. **42**(1), 5–30 (2011)
21. Fodor, J.A.: The Language of Thought. Thomas Y. Crowell, New York (1975)
22. Fodor, J.A., Pylyshyn, Z.W.: Connectionism and cognitive architecture: a critical analysis. Cognition **28**, 3–71 (1988)
23. d'Avila Garcez, A., Gori, M., Lamb, L.C., Serafini, L., Spranger, M., Tran, S.N.: Neural-symbolic computing: an effective methodology for principled integration of machine learning and reasoning. FLAP **6**(4), 611–632 (2019). https://collegepublications.co.uk/ifcolog/?00033
24. van Gelder, T.: The dynamical hypothesis in cognitive science. Behav. Brain Sci. **21**(5), 615–628 (1998)
25. Gomila, T., Calvo, P.: Directions for an embodied cognitive science: toward an integrated approach. In: Handbook of Cognitive Science, pp. 1–25. Elsevier (2008)
26. Goschke, T., Koppelberg, D.: Connectionist representation, semantic compositionality, and the instability of concept structure. Psychol. Res. **52**, 253–270 (1990)
27. Graves, A., et al.: Hybrid computing using a neural network with dynamic external memory. Nature **538**(7626), 471–476 (2016). http://dx.doi.org/10.1038/nature20101
28. Harman, G.: Conceptual role semantics. Notre Dame J. Formal Logic **28**, 252–256 (1982)
29. Harnad, S.: The symbol grounding problem. Physica D **42**, 335–346 (1990)

30. Hassabis, D., Kumaran, D., Summerfield, C., Botvinick, M.: Neuroscience-inspired artificial intelligence. Neuron **95**, 245–258 (2017). https://doi.org/10.1016/j.neuron.2017.06.011
31. Haugeland, J.: Artificial Intelligence: The Very Idea. Massachusetts Institute of Technology, USA (1985)
32. Hawkins, J., Blakeslee, S.: On Intelligence. Times Books, New York (2004)
33. Hayes, P.J.: The naïve physics manifesto. In: Michie, D. (ed.) Expert Systems in the Micro-Electronic Age, pp. 242–270. Edinburgh University Press, Edinburgh (1979)
34. Hinton, G.E., McClelland, J.L., Rumelhart, D.E.: Distributed representation. In: Rumelhart, D.E., McClelland, J.L. (eds.) Parallel Distributed Processing: Exploration in the Microstructure of cognition, Vol. 1, Foundations, pp. 77–109. MIT Press, Cambridge (1986)
35. Hofstadter, D., Sander, E.: Surfaces and Essences: Analogy as the Fuel and Fire of Thinking. Basic Books (2013)
36. Hofstadter, D.R.: The copycat project: an experiment in nondeterminism and creative analogies. AI memo, MIT Artificial Intelligence Laboratory (1984)
37. Hofstadter, D.R.: Waking up from the Boolean dream, or, subcognition as computation. In: Metamagical Themas: Questing for the Essence of Mind and Pattern, chap. 26. Basic Books, New York (1985)
38. Hofstadter, D.R.: On seeing A's and seeing As. Stanf. Humanit. Rev. **4**, 109–121 (1995)
39. Holland, J.H.: Escaping brittleness: the possibilities of general purpose learning algorithms applied to parallel rule-based systems. In: Michalski, R.S., Carbonell, J.G., Mitchell, T.M. (eds.) Machine Learning: An Artificial Intelligence Approach, vol. II, pp. 593–624. Morgan Kaufmann, Los Altos (1986)
40. Kneale, W., Kneale, M.: The Development of Logic. Clarendon Press, Oxford (1962)
41. Laird, J.E.: The Soar Cognitive Architecture. MIT Press, Cambridge (2012)
42. Lakoff, G.: Cognitive semantics. In: Eco, U., Santambrogio, M., Violi, P. (eds.) Meaning and Mental Representation, pp. 119–154. Indiana University Press, Bloomington (1988)
43. Laurence, S., Margolis, E.: Concepts and cognitive science. In: Margolis, E., Laurence, S. (eds.) Concepts: Core Readings, pp. 3–81. MIT Press, Cambridge (1999)
44. LeCun, Y., Bengio, Y., Hinton, G.: Deep learning. Nature **521**, 436–444 (2015). https://doi.org/10.1038/nature14539
45. Lenat, D.B., Feigenbaum, E.A.: On the thresholds of knowledge. Artif. Intell. **47**, 185–250 (1991)
46. Mach, E.: The Science of Mechanics. Read Books (2008)
47. Marcus, G.: The next decade in AI: four steps towards robust artificial intelligence (2020)
48. McCarthy, J.: Artificial intelligence, logic and formalizing common sense. In: Thomason, R.H. (ed.) Philosophical Logic and Artificial Intelligence, pp. 161–190. Kluwer, Dordrecht (1989)
49. McCarthy, J.: Notes on formalizing contexts. In: Proceedings of the Thirteenth International Joint Conference on Artificial Intelligence, pp. 555–560 (1993)
50. McCulloch, W.S., Pitts, W.H.: A logical calculus of ideas immanent in neural activity. Bull. Math. Biophys. **5**, 115–133 (1943)
51. McDermott, D.: A critique of pure reason. Comput. Intell. **3**, 151–160 (1987)
52. Minsky, M.: Neural nets and the brain-model problem. Ph.D. thesis, Department of Computer Science, Princeton University (1954)

53. Minsky, M.: A framework for representing knowledge (1974). MIT-AI Laboratory Memo No. 306
54. Minsky, M.: The Emotion Machine: Commonsense Thinking, Artificial Intelligence, and the Future of the Human Mind. Simon & Schuster (2006)
55. Newell, A.: The knowledge level: presidential address. AI Mag. **2**(2), 1–20 (1981)
56. Newell, A., Simon, H.A.: Computer science as empirical inquiry: symbols and search. Commun. ACM **19**(3), 113–126 (1976)
57. Nguyen, A., Yosinski, J., Clune, J.: Deep neural networks are easily fooled: high confidence predictions for unrecognizable images. In: 2015 IEEE Conference on Computer Vision and Pattern Recognition (2015)
58. Nivel, E., Thórisson, K.: Self-programming: operationalizing autonomy. In: Proceedings of the Second Conference on Artificial General Intelligence, pp. 150–155 (2009)
59. Nivel, E., Thórisson, K.: Towards a programming paradigm for control systems with high levels of existential autonomy. In: Proceedings of the Sixth Conference on Artificial General Intelligence, pp. 78–87 (2013)
60. Nivel, E., et al.: Autocatalytic Endogenous Reflective Architecture. Reykjavik University School of Computer Science Technical Report, RUTR-SCS13002 (2013). https://alumni.media.mit.edu/~kris/ftp/AERA-RUTR-SCS13002.pdf
61. Nivel, E., Thórisson, K.R., Steunebrink, B., Schmidhuber, J.: Anytime bounded rationality. In: Bieger, J., Goertzel, B., Potapov, A. (eds.) Proceedings of the 8th Conference on Artificial General Intelligence, pp. 121–130. Springer, Cham (2015)
62. Nivel, E., et al.: Bounded recursive self-improvement. CoRR abs/1312.6764 (2013). http://arxiv.org/abs/1312.6764
63. Ogden, C., Richard, I.: The Meaning of Meaning (10th edition, 1949). Kegan Paul, London (1923)
64. Pattee, H.H.: Cell psychology: an evolutionary approach to the symbol-matter problem. Cogn. Brain Theory **5**, 325–341 (1983)
65. Pearl, J., Mackenzie, D.: The Book of Why. Basic Books, New York (2018)
66. Piaget, J.: The Construction of Reality in the Child. Basic Books, New York (1954)
67. Piccinini, G.: The first computational theory of mind and brain: a close look at McCulloch and Pitts's 'logical calculus of ideas immanent in nervous activity'. Synthese **141**, 175–215 (2004)
68. Pinker, S., Prince, A.: On language and connectionism: analysis of a parallel distributed processing model of language acquisition. Cognition **28**, 73–193 (1988)
69. Pollock, J.: OSCAR: an architecture for generally intelligent agents. In: Proceedings of the First Conference on Artificial General Intelligence, pp. 275–286 (2008)
70. Pollock, J.L.: Thinking About Acting: Logical Foundations for Rational Decision Making. Oxford University Press, New York (2006)
71. Rogers, T.T., McClelland, J.L.: Précis of semantic cognition: a parallel distributed processing approach. Behav. Brain Sci. **31**, 689–749 (2008)
72. Rosenblatt, F.: The perceptron: a perceiving and recognizing automaton (1957). Cornell Aeronautical Laboratory Report 85-60-1
73. Rosenbloom, P.S., Demski, A., Ustun, V.: The sigma cognitive architecture and system: towards functionally elegant grand unification. J. Artif. Gen. Intell. **7**(1), 1–103 (2016)
74. Rumelhart, D.E., McClelland, J.L. (eds.): Parallel Distributed Processing: Explorations in the Microstructure of Cognition, vol. 1. Foundations. MIT Press, Cambridge (1986)

75. Schmidhuber, J.: Deep learning in neural networks: an overview. Neural Netw. **61**, 85–117 (2015)
76. Searle, J.: Minds, brains, and programs. Behav. Brain Sci. **3**, 417–424 (1980)
77. Sejnowski, T.J.: The unreasonable effectiveness of deep learning in artificial intelligence. Proc. Natl. Acad. Sci. (2020). https://doi.org/10.1073/pnas.1907373117
78. Sheikhlar, A., Thórisson, K.R.: Causal generalization via goal-driven analogy. In: Thórisson, K.R., Isaev, P., Sheikhlar, A. (eds.) International Conference on Artificial General Intelligence, pp. 165–175. Springer (2024)
79. Skinner, B.F.: About Behaviorism. Knopf Doubleday Publishing Group, New York (1974)
80. Smolensky, P.: On the proper treatment of connectionism. Behav. Brain Sci. **11**, 1–74 (1988)
81. Sun, R.: Robust reasoning: integrating rule-based and similarity-based reasoning. Artif. Intell. **75**, 241–295 (1995)
82. Thórisson, K.R.: A new constructivist AI: from manual methods to self-constructive systems. In: Wang, P., Goertzel, B. (eds.) Theoretical Foundations of Artificial General Intelligence, pp. 145–171. Atlantis Press, Paris (2012)
83. Thórisson, K.R.: Discretionarily constrained adaptation under insufficient knowledge & resources. J. Artif. Gen. Intell. **11**(2), 7–12 (2020)
84. Thórisson, K.R.: Seed-programmed autonomous general learning. Proc. Mach. Learn. Res. **131**, 32–70 (2020)
85. Thórisson, K.R., Bieger, J., Li, X., Wang, P.: Cumulative learning. In: Proceedings of the 12th International Conference on Artificial General Intelligence, pp. 198–208. Springer (2019)
86. Thórisson, K.R., Helgasson, H.P.: Cognitive architectures and autonomy: a comparative review. J. Artif. Gen. Intell. **3**(2), 1–30 (2012)
87. Thórisson, K.R., Kremelberg, D., Steunebrink, B.R., Nivel, E.: About understanding. In: The Proceedings of the Ninth Conference on Artificial General Intelligence, pp. 106–117 (2016)
88. Thórisson, K.R., Minsky, H.: The future of AI research: ten defeasible 'axioms of intelligence'. Proc. Mach. Learn. Res. **192**, 5–21 (2023)
89. Thórisson, K.R., Nivel, E., Sanz, R., Wang, P.: Approaches and assumptions of self-programming in achieving artificial general intelligence. J. Artif. Gen. Intell. **3**(3), 1–10 (2012)
90. Thórisson, K.R., et al.: Autonomous acquisition of situated natural communication. Comput. Sci. Inf. Syst. **9**(2), 115–131 (2014)
91. Thórisson, K.R., Talbot, A.: Cumulative learning with causal-relational models. In: Artificial General Intelligence, pp. 227–237. Springer, Cham (2018)
92. Thórisson, K.R., Talevi, G.: A theory of foundational meaning generation in autonomous systems, natural & artificial. In: Thórisson, K.R., Isaev, P., Sheikhlar, A. (eds.) International Conference on Artificial General Intelligence, pp. 188–198. Springer (2024)
93. Wang, P.: Non-Axiomatic Reasoning System: Exploring the Essence of Intelligence. Ph.D. thesis, Indiana University (1995)
94. Wang, P.: The logic of learning. In: Working Notes of the AAAI Workshop on New Research Problems for Machine Learning, Austin, Texas, pp. 37–40 (2000)
95. Wang, P.: Experience-grounded semantics: a theory for intelligent systems. Cogn. Syst. Res. **6**(4), 282–302 (2005)
96. Wang, P.: Artificial general intelligence and classical neural network. In: Proceedings of the IEEE International Conference on Granular Computing, Atlanta, Georgia (2006)

97. Wang, P.: Rigid Flexibility: The Logic of Intelligence. Springer, Dordrecht (2006)
98. Wang, P.: Three fundamental misconceptions of artificial intelligence. J. Exp. Theor. Artif. Intell. **19**(3), 249–268 (2007)
99. Wang, P.: Case-by-case problem solving. In: Goertzel, B., Hitzler, P., Hutter, M. (eds.) Proceedings of the Second Conference on Artificial General Intelligence, pp. 180–185 (2009)
100. Wang, P.: Embodiment: does a laptop have a body? In: Goertzel, B., Hitzler, P., Hutter, M. (eds.) Proceedings of the Second Conference on Artificial General Intelligence, pp. 174–179 (2009)
101. Wang, P.: The assumptions on knowledge and resources in models of rationality. Int. J. Mach. Conscious. **3**(1), 193–218 (2011)
102. Wang, P.: Motivation management in AGI systems. In: Bach, J., Goertzel, B., Iklé, M. (eds.) Proceedings of the Fifth Conference on Artificial General Intelligence, pp. 352–361 (2012)
103. Wang, P.: Non-Axiomatic Logic: A Model of Intelligent Reasoning. World Scientific, Singapore (2013)
104. Wang, P.: On defining artificial intelligence. J. Artif. Gen. Intell. **10**(2), 1–37 (2019). https://doi.org/10.2478/jagi-2019-0002
105. Wang, P.: Toward a logic of everyday reasoning. In: Vallverdú, J., Müller, V.C. (eds.) Blended Cognition: The Robotic Challenge, pp. 275–302. Springer, Cham (2019). https://doi.org/10.1007/978-3-030-03104-6_11
106. Wang, P., Hofstadter, D.: A logic of categorization. J. Exp. Theor. Artif. Intell. **18**(2), 193–213 (2006)
107. Wang, P., Li, X.: Different conceptions of learning: function approximation vs. self-organization. In: Steunebrink, B., Wang, P., Goertzel, B. (eds.) Proceedings of the Ninth Conference on Artificial General Intelligence, pp. 140–149 (2016)
108. Wang, P., Steunebrink, B., Thórisson, K.R.: What should AGI learn from AI and CogSci? In: Goertzel, B., Orseau, L., Snaider, J. (eds.) Presented at the International Conference on Artificial General Intelligence. Springer (2014)
109. Watson, J.B.: Psychology as the behaviorist views it. Psychol. Rev. **20**(2), 158–177 (1913)
110. Winograd, T.: Thinking machines: can there be? Are we? In: The Foundation of Artificial Intelligence—A Sourcebook, pp. 167–189. Cambridge University Press, USA (1990)
111. Wittgenstein, L.: Philosophical Investigations. Prentice Hall, Upper Saddle River (1999). Translated by G. Anscombe
112. Woods, W.A.: What's in a link: foundations for semantic networks. In: Bobrow, D.G., Collins, A.M. (eds.) Representation and Understanding: Studies in Cognitive Science, pp. 35–82. Academic Press, New York (1975)

Society of LLMs: Constructivist Learning Approaches with Multiple LLM Instances

Steve Kommrusch(✉) and Henry Minsky

Leela AI, Somerville, USA
{sk,hqm}@leela-ai.com

Abstract. Large language models have been able to talk with some fluidity, have been shown to possess a rudimentary world model, and have some simple abilities to reason given careful prompting. In parallel, self-situated learning techniques have shown that machines can learn to build hypotheses about the world and discover new correlations through experimentation. We propose a system which interconnects LLMs in such a way that learning from experience is enabled. We propose using an open-source LLM (such as Llama 3.1 70b) which allows for incremental model training and have potentially over a dozen instances interacting together. We reference AutoGPT, AgentCoder, Curiosity Driven Red Teaming, and other such systems to illustrate how LLMs can be effectively interconnected. Our proposed system mimics many of the features of Gary Drescher's schema mechanism based on Piaget's work: one agent will propose goals to satisfy curiosity-based exploration; multiple agents will propose plans; when the plans differ but one succeeds in reaching a goal, an agent will create a new training sample; one agent may be 'thinking subconsciously' to create context other agents may use in a manner not unlike sub-activation, etc. Iterative training will be used to improve the single model as new hypothesis about the world are discovered.

Keywords: Machine Learning · Artificial Intelligence · Constructivism

1 Introduction

Modern LLMs, such as GPT-4 and Llama 3.1, exhibit fluid conversation, contextual understanding, and limited reasoning through prompt engineering. However, these models lack self-generated learning mechanisms that allow them to form and refine world models through experience. Constructivist AI, drawing from Jean Piaget's cognitive development theories, suggests that learning occurs through hypothesis generation, experimentation, and hypothesis refinement. We propose a multi-agent LLM system that facilitates self-improvement by enabling inter-agent interactions, goal-oriented exploration, and incremental model updates.

P. Robertson and O. Georgeon (Eds.): IWSSL 2024, LNCS 15457, pp. 53–61, 2025.
https://doi.org/10.1007/978-3-031-96325-4_3

2 Related Work

2.1 Large Language Models and World Models

LLMs have demonstrated abilities such as in-context learning, reasoning via chain-of-thought prompting, and task-specific fine-tuning [1]. Prior research suggests that transformer-based architectures can approximate world models to an extent, though they lack autonomous hypothesis formation and empirical validation (Fig. 1).

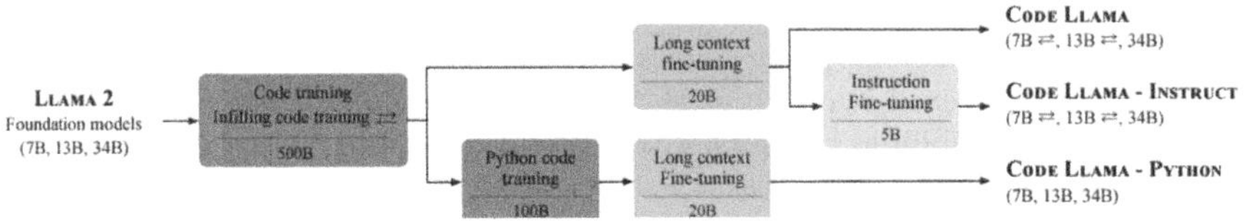

Fig. 1. Code Llama begins with an open-source large language model and performs fine-tuning on the model weights to improve its ability to generate code.

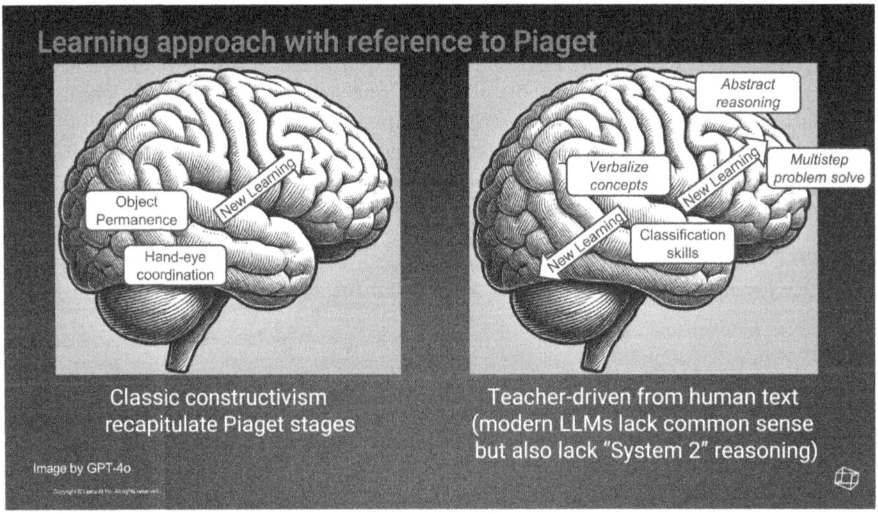

Fig. 2. Some early attempts at modeling human learning began with a blank slated and attempted to build world models using low-level constructivist concepts. We propose starting with the success of LLMs, but using world-exploration and online learning concepts from constructivist learning to improve low-level common sense and high-level abstract reasoning.

2.2 Constructivist Learning and Symbolic AI

Constructivism, pioneered by Jean Piaget, emphasizes knowledge construction through active exploration. Gary Drescher's schema mechanism [2] further formalizes this into computational models, enabling AI to build structured knowledge representations. Our proposed system leverages constructivist symbolic learning techniques into an LLM-driven architecture, leveraging self-supervised interactions to refine the model dynamically (Fig. 2).

2.3 Multi-agent LLM Systems

Approaches such as AutoGPT [3], AgentCoder [4] (Fig. 3), and Curiosity-Driven Red Teaming [5] have explored self-improving LLM interactions. These systems demonstrate how multiple instances can collaborate, test hypotheses, and refine knowledge. Our framework extends these approaches by incorporating goal-oriented planning, contextual sub-activation, and experience-based reinforcement.

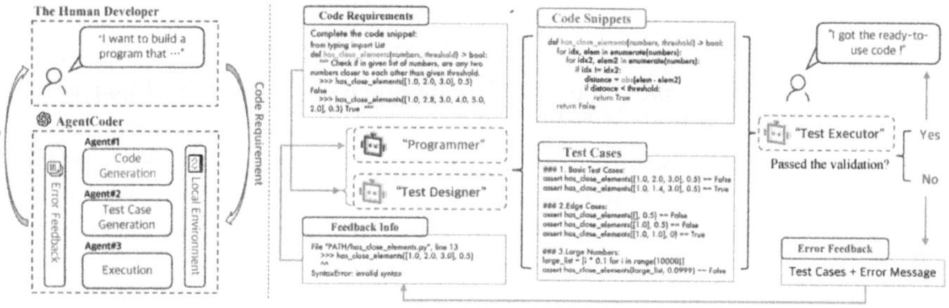

Fig. 3. Multiple instances of the same LLM collaborate to solve coding problems. One instance write code, a second writes tests, and a third evaluates test results and proposes test or code updates.

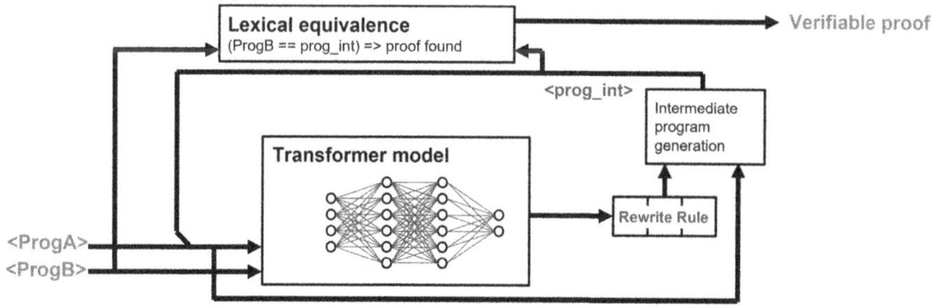

Fig. 4. A neural network is trained to find heuristics to help solve the NP-hard problem of program equivalence. Verifying the proof takes polynomial time, allowing verified correct results to be identified for iteratively improving the model.

2.4 Iterative Self-supervised Learning

Systems such as S4Eq [6] (Fig. 4) show how a transformer model can be iteratively trained based on verifiable outputs to questions beyond the original training dataset. The key to such systems is having a robust mechanism to verify the correctness of the model output.

3 Proposed System Architecture

We propose an LLM-based constructivist learning system where multiple instances interact to refine world models and generate new knowledge. The system consists of the following core components:

- Base action models instance
- Curiosity-Driven Goal Setting and Hypothesis Generation
- Multi-agent planning
- Training Sample Generation with Alignment Goals
- Contextual Sub-Activation
- Incremental Model Update

The key idea we propose is that each LLM in the society is prompted differently such that its compartmentalized behavior is performed and learned. One could then have the different LLM models learn independently, but we hypothesize that incrementally training a single LLM model, given different prompts, will help the LLM model generalize better as a broader set of incremental learning is provided compared to allowing the LLMs to diverge during training.

3.1 Base Action Model Instance

When given a goal, the base action model instance will be prompted to create a plan to reach the goal and then to create 2 or more alternate plans that may work better or help with recovery if the first fails. By providing the initial 'best guess' and 2 alternates, we can use the order to identify 'surprise' results and train the model to prefer an alternate plan in the future if further analysis warrants.

3.2 Curiosity-Driven Goal Setting and Hypothesis Generation

If a user interacts with the Society of LLMs, they can provide input which becomes a goal to solve. In other cases, one agent generates exploratory goals to guide learning. Goals are formulated based on uncertainty, novelty, and prior learning gaps. Hence, the curiosity LLM would have as input context the recent goals, user queries, and actionable outputs. Its prompt would encourage reducing uncertainty about the current world state or improving prediction ability if there seem to be multiple possible results from an action. Additionally, the model can be prompted to explore some action if it hypothesizes that the action could provide information useful for reaching the goal.

3.3 Multi-agent Planning

Once a plan of action exists, multiple agents can probe the plan for soundness. One agent may analyze the plan for safety concerns (such as unintended side effects), another may try to reverse the plan and reason about the context needed to take each action (to 'look at the problem from a different angle'), etc. The base action model instance will have created multiple plans and could be asked to create more if this analysis step fails to accept any of the plans for execution.

3.4 Training Sample Generation with Alignment Goals

After a plan is attempted in the world, its results are evaluated for possible learning. If the original initial plan from the base model succeeds, in general no interesting learning occurred. If a secondary plan was attempted after analysis and it succeeded, then multiple training samples may be generated: one sample from the successful secondary plan, but also a sample for the multi-agent planner which selected the secondary plan over the original plan. Additionally, even when a plan fails, the result may be of interest. For example, in 'hindsight' learning, the resulting world state might be an interesting future goal, so a training sample can be created that modifies the query to be the goal that resulted and then the action plan is correct. A properly prompted LLM does the results analysis for sample selection. Sample selection should include prompting to analyze new samples for alignment concerns as incremental training may effect the inner alignment of a model over time. Iterative fine-tuning using LoRA can be done for every plan that is attempted.

3.5 Contextual Sub-activation

One agent acts as a subconscious thinker, continuously analyzing patterns. Sub-activation allows for holes or even inconsistencies in the LLM's world model to be addressed. As in Gary Drescher's work, this process involves considering a goal state and how it might be achieved but not taking action to verify the result. Instead, if the 'multi-agent planning' process can reach a confidence threshold on the result of an action sequence, and the action sequence itself is 'surprising' in some way, then the sequence may be added to incremental learning on the LLM as if it had been tested in the world.

3.6 Incremental Model Updates

The LLM undergoes continuous fine-tuning based on validated discoveries, as well as hindsight learning when relevant. Model refinement occurs through self-supervised, iterative learning cycles. Temporary LoRA model adjustments can be used to test if a model's update will improve future predictions about the current environment before applying the adjustments to the actual model weights.

4 Experimental Setup and Evaluation Metrics

4.1 Experimental Design

Our broad proposal could apply to a variety of environments on which actions can be taken, including robot actions on the real world. However, for demonstration we plan to focus on the ARC-AGI challenge as it has been shown to be a challenging problem that we believe our approach can make demonstrable progress on. Figure 5 shows the input and output expectations for this problem. We propose evaluating our system using:

- Base model is prompted with ARC demonstration pairs and is asked to generate python code which receives the inputs and generates the outputs. Multiple program proposals would be generated.
- Curiosity and hypothesis generation would be modeled with an LLM instance that tests the code on each input from the input/output demonstration pairs. Given 3 or 4 challenges in the LLM context buffer, new python subroutines could be proposed by the model which could be used for multiple problems. An LLM could be tasked with generating augmented training data which is similar to the original but modified to explore a broader distribution for the learning step.
- Multi-agent planning would still evaluate which of the proposed python programs is best for creating an ARC answer. This step would including executing the proposed programs and providing feedback to the base model, similar to AgentCoder.
- Incremental Model Updates We would use LoRA to provide incremental updates to the model, but add the updated weights to the base parameters (instead of holding them out for test-time compute).

Figure 6 shows conceptually how a Society of LLMs can be structured for a problem like the ARC-AGI challenge. The figure shows multiple LLM agents engaged in solving a coding problem while other agents are engaged in sample analysis for incremental training.

4.2 Evaluation Metrics

We would measure our success by observing our score on the ARC challenge. We would test our original LLM instance, then our incrementally trained models. Comparing to other models of similar size in the literature on this problem would allow us to determine the value of our approach.

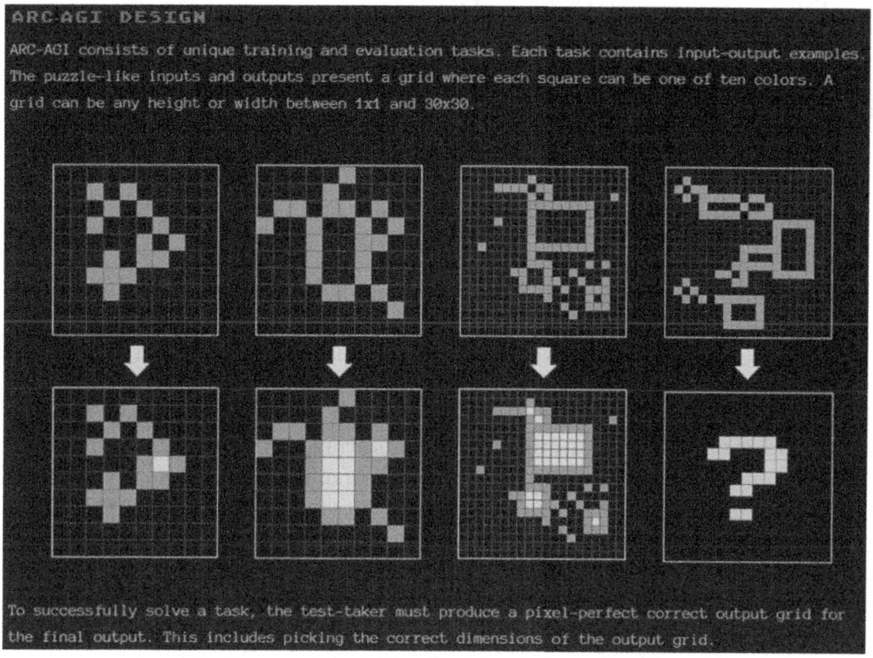

Fig. 5. The Abstraction and Reasoning Corpus (ARC) includes demonstration pairs and a test input for which the AI (or human) should create the output. Understanding the abstract concepts illustrated in the demonstration pairs allows for correct output prediction.

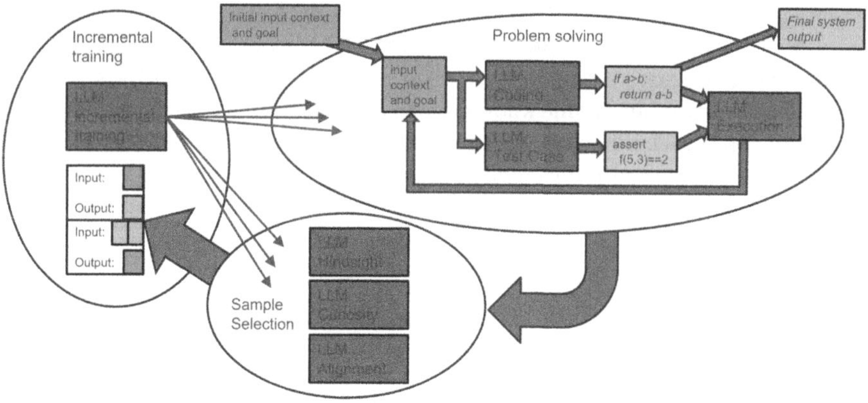

Fig. 6. In our proposal, a single LLM model (indicated as orange boxes) is given various prompts to perform different roles in the LLM society. The yellow arrows indicate scaffolding connections which provide for interaction among LLM models. After problem solving attempts, incremental learning of the model can be done using selected data, thus allowing the LLM model to learn through experimentation in the world it acts upon. (Color figure online)

5 Discussion and Future Directions

Our proposed system bridges the gap between static, pre-trained LLMs and dynamic, self-improving AI. We highlight several key areas for future research:

- Enhancing multi-agent collaboration mechanisms for better hypothesis validation.
- Developing scalable self-supervised training pipelines for real-world applications.
- Extending constructivist principles to incremental learning for autonomous adaptation.
- Understand ability of selecting for aligned incremental learning training samples to improve alignment of actions taken [7].

We believe that this approach will significantly improve LLM reasoning, learning efficiency, and contextual awareness, paving the way for more adaptive, AGI-like systems.

Using multiple instances of LLMs can increase computational inference costs. We propose studying the ability of a small or modest size model to improve using our iterative learning approach. If such models can be trained to perform better than large models on certain reasoning tasks, that would be a strong success for our approach but also allow for smaller models to be used where large but less capable models had been used before.

6 Conclusion

By integrating constructivist learning principles with multi-agent LLM architectures, we propose a novel methodology for continuous AI self-improvement. Inspired by Drescher's schema mechanism, our system enables LLMs to iteratively generate, test, and refine knowledge in a structured manner. This approach has significant implications for self-improving AI, automated research assistants, and next-generation reasoning models.

References

1. Rozière, B., et al.: Code Llama: Open Foundation Models for Code. arXiv e-prints (2024). https://arxiv.org/abs/2308.12950
2. Drescher, G: Made Up Minds. The MIT Press (1991)
3. Ortiz, S.: What is AutoGPT? Everything to know about the next powerful AI tool. ZDNET. Accessed 16 Apr 2023
4. Huang, D., Bu, Q., Zhang, J., Luck, M., Cui, H.: AgentCoder: Multiagent-Code Generation with Iterative Testing and Optimisation. arXiv e-prints (2024). https://arxiv.org/html/2312.13010v2
5. Hong, Z.-W., et al.: Curiosity-driven Red-teaming for Large Language Models. arxiv e-prints (2024). https://arxiv.org/abs/2402.19464

6. Kommrusch, S., Monperrus, M., Pouchet, L.-N.: Self-supervised learning to prove equivalence between straight-line programs via rewrite rules. IEEE Trans. Softw. Eng. 3771–3792 (2023)
7. Lehalleur, S.P., et al.: You Are What You Eat - AI Alignment Requires Understanding How Data Shapes Structure and Generalisation. arxiv e-prints (2025). https://arxiv.org/abs/2502.05475v1

Schema Mechanisms 2.0 for Developmental Artificial Intelligence

Olivier L. Georgeon[1(✉)], Filipo S. Perotto[2], Kristinn R. Thórisson[3], Arash Sheikhlar[3], and Paul Robertson[4]

[1] UR CONFLUENCE: Sciences et Humanites (EA 1598), UCLy, Lyon, France
ogeorgeon@univ-catholyon.fr
[2] ONERA – The French Aerospace Lab – DTIS, Toulouse, France
filipo.perotto@onera.fr
[3] Center for Analysis and Design of Intelligent Agents, Reykjavik University, Reykjavik, Iceland
thorisson@ru.is, arash@iiim.is
[4] DOLL Labs, Lexington, MA, USA
paulr@dollabs.com

Abstract. Schema mechanisms are software frameworks designed to reflect learning theories like genetic epistemology, constructivist epistemology, and developmental learning. Building on Gary Drescher's pioneering implementation from 1991, this paper reviews four contemporary schema mechanisms: CALM, LIDA, ECA, and AERA. While maintaining Drescher's Piagetian-inspired view that sensorimotor schemas serve as the foundation for knowledge construction, these new mechanisms extend the constructivist approach by grounding schemas in interactional events rather than static world perceptions. This shift introduces an interactional motivation framework and integrates hierarchical abstraction, internal simulation, and physical space into schemas, paving the way for their application in autonomous robotics operating in open environments. We believe that these advancements mark the emergence of a new generation of schema mechanisms, which we refer to as Schema Mechanisms 2.0.

Keywords: schema mechanism · genetic epistemology · constructivist learning · intrinsic motivation · cognitive architectures

1 Introduction

Throughout his life, Jean Piaget developed and popularized the theory of *genetic epistemology* [38] to account for the genesis of intelligence and knowledge. In parallel, he pioneered *developmental psychology* by establishing the foundational methods for studying mental development in children [36,37]. Toward the end of his life, he connected genetic epistemology with constructivist epistemology and the work of Ernst von Glasersfeld [18]. Genetic epistemology rests on the key concept of "schema", closely related to the concept of "functional circles" proposed by Jakob von Uexküll. We refer to Ziemke [53] for a broader discussion

P. Robertson and O. Georgeon (Eds.): IWSSL 2024, LNCS 15457, pp. 62–87, 2025.
https://doi.org/10.1007/978-3-031-96325-4_4

on the philosophical roots of constructivist epistemology in Kant's philosophy, and Uexküll's biology. Piaget defines a schema as follows:

> "A schema is a structure or organization of actions as they are transferred or generalized by repetition in similar or analogous circumstances." ([35], p. 4).

A schema is the basic unit of knowledge that encapsulates an action and its circumstances, that is, a *pattern of interaction.* Genetic epistemology insists on the primacy of interaction as a condition for the emergence of perception and knowledge, surpassing innatist and empiricist explanations:

> "Knowledge does not originally arise either from a subject conscious of itself or from objects already constituted (from the subject's point of view) that would impose themselves on the subject. Knowledge results from interactions occurring halfway between the subject and the objects, and thus involving both, but due to a complete un-differentiation and not from exchanges between distinct forms.
>
> If, at the beginning, there is neither a subject, in the epistemic sense of the term, nor objects, conceived as such, nor, above all, invariant instruments of exchange, then the initial problem of knowledge will be to construct such mediators. Starting from the contact zone between one's own body and the objects, these mediators will progressively engage more deeply in both complementary directions toward the exterior and the interior. It is from this dual progressive construction that the joint elaboration of both the subject and the objects depends.
>
> The initial instrument of exchange is not perception, as rationalists too easily conceded to empiricism, but rather action itself, with its much greater plasticity. Certainly, perceptions play an essential role, but they partly depend on action as a whole, and some perceptual mechanisms that one might have thought to be innate or very primitive only emerge at a certain level of object construction." (translated from [39], p. 14-15)

For Piaget, intelligence is related to the ability to gradually organize and complexify the set of schemas. Increasing organization of internal structures are guided by two forces: by assimilating new experiences to existing schemas, the mind tries to integrate observations and interactions in terms of already constructed cognitive structures, extending or generalizing their scope, and by accommodating its set of schemas to surprising or disruptive experiences, differentiating, specializing, complexifying its internal relations, the mind creates new intellectual resources for better understanding and mastering its interactions:

> "[...] Organization is inseparable from adaptation: they are two complementary processes of a single mechanism, organization being the internal aspect of the cycle of which adaptation constitutes the external aspect. [...] that adaptation is an equilibrium between assimilation and accommodation. [...] Intelligence is assimilation to the extent that it incorporates all the given data of experience within its framework. [...] That mental life is also accommodation to the environment. Assimilation can never be pure because by incorporating new elements into its earlier schemata the intelligence constantly modifies the latter in order to adjust them to new elements." ([37], p. 7)
>
> "In their initial directions, assimilation and accommodation are obviously opposed to one another, since assimilation is conservative and tends to subordinate the environment to the organism as it is, whereas accommodation is the source of changes and bends the organism to the successive constraints of the environment. But if in their rudiment these two functions are antagonistic, it is precisely the role of mental life in general and of intelligence in particular to intercoordinate them." ([36], p. 352)
>
> "Gradually, as the child's thought evolves, assimilation and accommodation are differentiated and become increasingly complementary. In the realm of representation of the world this means, on the one hand, that accommodation, instead of remaining on the surface of experience, penetrates it more and more deeply, that is, under the chaos of appearances it seeks regularities and becomes capable of real experimentations to establish them. On the other hand, assimilation, instead of reducing phenomena to the concepts inspired by personal activity, incorporates them in the system of relationships rising from the more profound activity of intelligence itself." ([36], p. 385)
>
> "The study of sensorimotor or practical intelligence in the first two years of development has taught us how the child, at first directly assimilating the external environment to his own activity, later, in order to extend this assimilation, forms an increasing number of schemata which are both more mobile and better able to intercoordinate. [...] The increasing coherence of the schemata thus parallels the formation of a world of objects and spatial relationships, in short, the elaboration of a solid and permanent universe." ([36], p.*xi*)

Schema mechanisms attempt to reflect the key insights of genetic epistemology effectively expressed by these words of Piaget regarding the primary role of sensorimotor experience and organization of mental structures. Indeed, genetic epistemology appears to provide a materialistic explanation of how the human mind develops from birth, potentially lending itself to computational implementation.

Over the past decades, Piaget's theories have remained central to contemporary theories of knowledge development, continually questioned and shaped by active debates. Some shortcomings have been identified, including an underestimation of infants' abilities and an insufficient consideration for cultural and social factors (e.g., [1]). The relationship between constructivist and nativist theories remains a topic of debate, raising questions about the innate knowledge and prerequisites that should be incorporated into any computational implementation [5, p. 41]. The ongoing research on schema mechanisms examined in this paper contributes to these debates.

2 Schema Mechanisms

Implementing genetic epistemology as a mechanistic process involves reducing Piaget's psychological concept of schema into data structures handled by a computer. This computer controls an artifact (robot or virtual agent) that interacts with its environment. The computer must process these data structures through some mechanism that will support the generation of increasingly intelligent behavior demonstrating the internal development of the robot. Guerin and McKenzie [19] proposed the graphical representation of this developmental process shown in the Fig. 1.

Notably, the developmental process is not targeted at reaching a predefined goal-state in a predefined problem space, nor does it aim at maximizing a predefined scalar value as in reinforcement learning. Instead, the problem of cognitive development in artificial systems is intertwined with the problem *intrinsic motivation*, as has been argued for example by Oudeyer and colleagues [28].

2.1 Drescher's Schema Mechanism

In the early 1990s, Gary Drescher pioneered the first schema mechanism as a result of his PhD thesis [4] and associated publications [3,6], which gave rise to a successful book [5], proposing a framework to reproduce key constructivist notions in artificial intelligence. Drescher formalized schemas as the tuple ⟨*context*, *action*, *result*⟩ depicted in Fig. 2. The *context* and the *result* are sets of binary *items* that can be *positively* or *negatively* included. A schema's context is satisfied when all the positively included items are *On* and all the negatively included items are *Off*. A schema's result asserts that, if the *action* is taken when the context is satisfied, then the positively included items of the results will be turned *On* and its negatively included items will be turned *Off*.

The agent is initialized with a set of *primitive* items and actions that represent builtin elementary sensorimotor variables. For each primitive action, a *bare schema* with empty context and result is created. The designer may also add other initial schemas that encode the agent's "innate" behaviors.

The *activation* of a schema consists of initiating its action. Upon completion of the action, the predicted result is compared with the actual result. The schema is said to *succeed* if its predicted results are all obtained, and to *fail* otherwise.

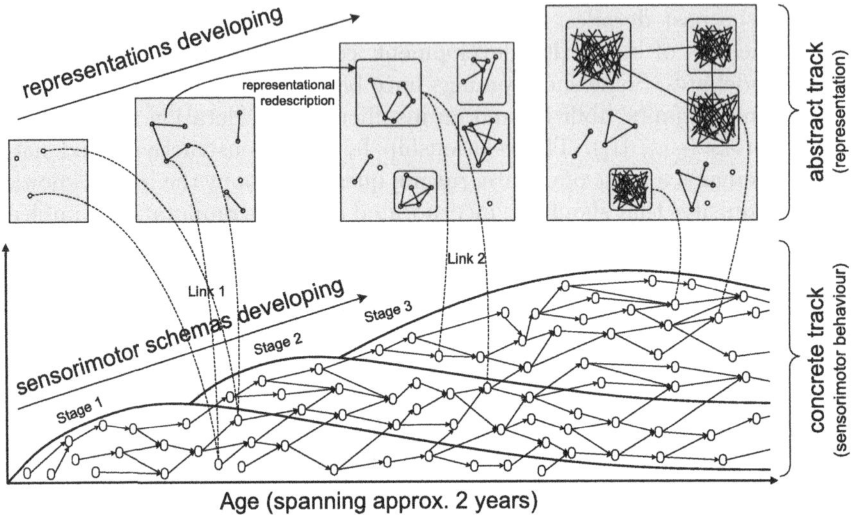

Fig. 1. Conceptual diagram of infant development from [19] Fig. 1. The lower (concrete) track shows a directed acyclic graph of sensorimotor schemas. A node represents a newly created schema. An edge has the meaning "is a necessary precursor". Stage 1: behaviors without objects. Stage 2: behaviors with single objects. Stage 3: object-object behaviors. The schemas now involve relationships among objects, and locations and transforms within space. The higher (abstract) track represents representations of objects by schemas and physical properties influencing their interactions.

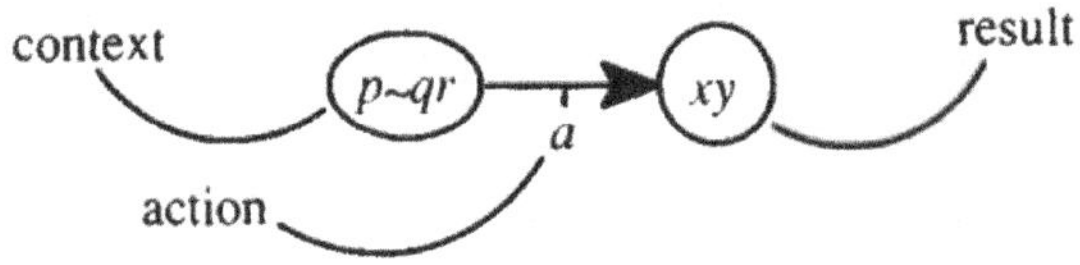

Fig. 2. Schema from [5] Fig. 3.2. The schema noted $p{\sim}qr/a/xy$ asserts that if action a is taken in the context where item p is On, q is Off, and r is On then the items x and y will be turned On.

The ratio of success and failure is called *reliability* and is memorized for each schema.

The learning of new schemas is triggered by the identification of *relevant results* made of items that are slightly more frequent than average. When a relevant result is identified, the agent seeks conditions under which it follows reliably. Such conditions are found through new schemas that allow recovering the context leading to the relevant result, and then performing the action again to asses the reliability of these new schemas.

After learning new schemas, the agent can learn new "concepts" when it identifies schemas that act as "probes" that may evoke the manifestation of a "thing". These schemas are only reliable if the "thing" is present. When such a schema is found, it is considered a *host schema* for a newly spawned *synthetic item* that represents this thing. Eventually, the agent finds out that a single "thing" may manifest itself through different host schemas. Drescher notes that "a synthetic item thus works backward from a thing's manifestation to define the very thing manifested" [5, p. 83]. This addresses the fundamental problem of *concept invention* in that "a synthetic item is a new element of the system's ontology—an element fundamentally different from the prior contents of the system's conceptual vocabulary" [5, p. 81].

The designer defines the agent's goals by assigning *primitive values* (positive if desirable, negative if dreaded) to values of items (*On* or *Off*). The mechanism can identify subgoals as specific sets of items to be activated or deactivated; and it constructs *composite actions* as sequences of schemas used to reach a subgoal. The values of a target item may be propagated to other items in the form of *instrumental values* and *delegated values* when attaining these other items improves the chances to reach the target item. The agent randomly alternates between explicit goal-pursuit and exploration, with probability defined by the designer.

Drescher demonstrates its schema mechanism in the "microworld" made of the two-dimensional grid shown in Fig. 3. "The mechanism controls a simulated robot that has a body, a single hand, and a visual system. The hand can touch and grasp objects, and move them about. The visual system maps a visual field onto a region of the world in the immediate vicinity of the robot body" [5, p. 114].

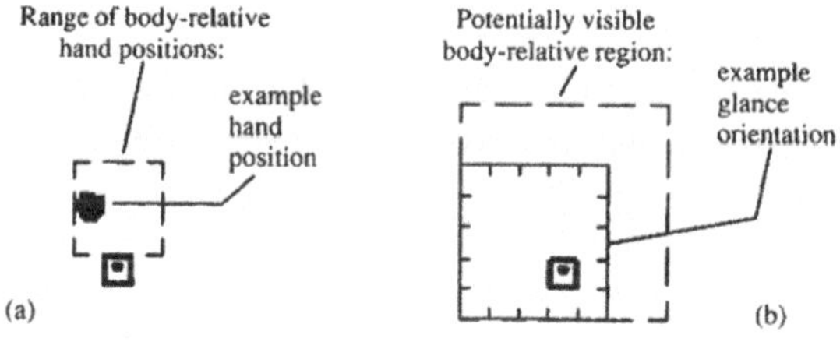

Fig. 3. Drescher's benchmark from [5, Fig. 6.1]. The simulated robot's body is represented by the small square with a black dot. (a): The hand (black spot) can move in the 3×3 grid in front of the body. (b): The visual field is the 5×5 grid in the environment made of a 7×7 grid.

The mechanism initially ignores the topological relations among sensory items, but the demonstration shows that it gradually learns these relations in a process that Drescher calls the "elaboration of the spatial substrate". Such learning of a cognitive map from sensorimotor sequences has recently been further investigated by Raju et al. [40].

Once the spatial substrate is learned, the mechanism manages to construct synthetic items that designate objects as distinct from their current perceptions. Drescher gives the example of "a synthetic item, which we might call

`PalpableObjectAt(1,3)` (the name, of course, has no meaning to the mechanism). The item is defined to represent whatever condition makes the schema reliable; we observers know (but the mechanism does not) that this condition is that there be an object at (1, 3)" [5, p. 13]. Synthetic item `PalpableObjectAt(1,3)` may later be associated with `VisibleObjectAt(1,3)` to account for the discovery that they refer to the same thing at position (1, 3) in the world.

In summary, Drescher's schema mechanism stands among the first attempts to investigate Piaget's question of how the child's mind can dynamically construct its own knowledge of reality from experience of interaction. In our view, Drescher's main contribution to this question rests on the construction of synthetic items to account for "things in the world" that the agent can discover through regularities in sequences of sensorimotor experience. Drescher goes beyond his predecessors (e.g., [20]) who merely categorized passively-received inputs. His schema mechanism was the first that "grounds its synthetic items in the reification of counterfactual assertions" [5, p. 90].

Drescher identified some limitations that may explain why nobody has made his mechanism work in a more complex environment since its creation almost 40 years ago. These limitations follow from the acknowledgment that "whatever experiences an organism or mechanism has had, there are infinitely many mutually contradictory generalizations that are consistent with those experiences. Therefore, any induction apparatus must impose a choice among those consistent generalizations" [5, p. 174].

Due to specific design choices, Drescher's schema mechanism may lack generality. For example the assumption that a sensory item "is a state element" and "each item represents some condition in the world" [5, p. 56] presupposes that the world is defined as a predefined set of states, which is acceptable in a microworld but difficult in the physical world. Another assumption was to model the agent's motivation through positive and negative values associated with sensory items. The exploration mechanism based on hysteresis and habituation may also lack generality; other motivational principles can be imagined. Conversely, a schema mechanism may need additional built-in assumptions to cope with the complexity of the physical world. This paper reviews other schema mechanisms that make other choices in the hope to move forward towards a general schema mechanism that could work in the real world.

2.2 Constructivist Anticipatory Learning Mechanism (CALM)

In the early 2010s, Filipo S. Perotto proposed his own version of a schema mechanism, called *Constructivist Anticipatory Learning Mechanism* (CALM), as a result of his PhD thesis [29,30] and related publications [31–34]. Perotto tackled the same challenges as Drescher, namely: learning a model of the regularities observed by an agent while interacting with the environment, and extending the representational vocabulary to allow the agent to overpass the raw sensorimotor perception by the creation of more abstract concepts. However, while Drescher's mechanism is closer to the 1980s symbolic planning languages and structures, Perotto's background lied on reinforcement learning [47] and multiagent systems.

In this way, CALM can be seen as a kind of model-based RL algorithm, trying to learn a model of the transitions (i.e. how the observations change depending on the context and actions), then, at each time step, computing a policy of actions that maximizes the expected discounted sum of rewards, similar to adaptive dynamic programming methods. Like Drescher's schema mechanism, the elementary knowledge structures in CALM are schemas, illustrated in the Fig. 4, represented by tuples associating context, action, and expected result.

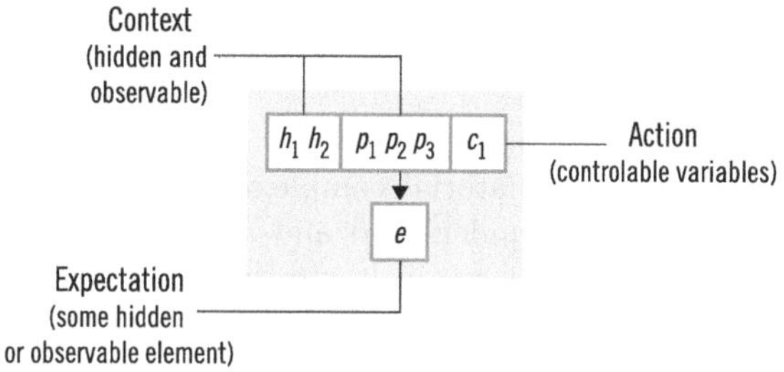

Fig. 4. Perotto's schema from [31] Fig. 2. A schema is composed of three vectors: context containing hidden variables and observable properties $(h_1, h_2, p_1, p_2, p_3)$, action (c_1), and expected result (e).

The main drawback is that CALM can only identify regularities if they are deterministic, differently from the original schema mechanism, which can maintain unreliable regularities within the system in order to spinoff reliable schemas afterwards. To compensate that limitation, CALM introduces a new distinction between two kinds of context variables: hidden variables and observable variables as shown in the Fig. 4. The justification is that some regularities appear as non-deterministic because the agent cannot perceive all the contextual elements necessary to precisely determine its transition function. In this way, the accent is placed on the discovering of abstract or non-observable elements, that once introduced and integrated in the anticipatory system, are able to explain the observation, making the transition deterministic. The idea is, in fact, quite similar to what Drescher called synthetic items in the original mechanism, but the way to create them is slightly different: when an expected result is contradicted, and no elements from context or action can explain the incoherence between the prediction and the observed result, then the mechanism supposes the existence of an abstract variable that, if known, would be able do distinguish between the two situations. The observed result then serves as a mean to deduce the value of that abstract variable, i.e. when the result corresponds to the old expectation, it is because the non-observable condition was present, and if the result is the new one, then the non-observable condition was absent, which corresponds to the state of the abstract variable perceived a posteriori. Finally, by searching in the previous step, the mechanism can try to anticipate the contexts and actions that changes that abstract variable, like illustrated in the Fig. 5.

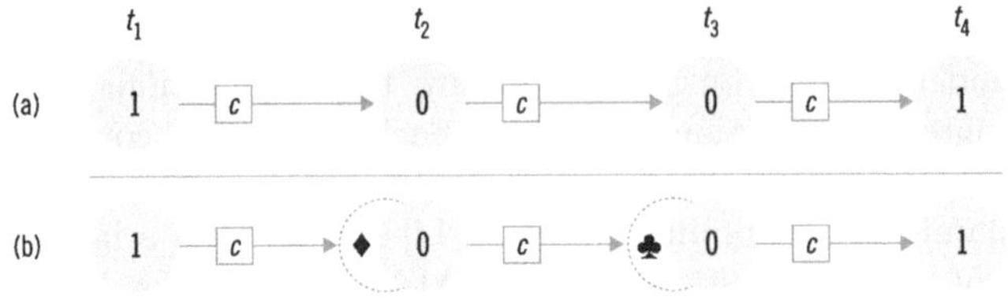

Fig. 5. Predicting the dynamics of a non-observable property. (a): An actually experienced sequence including $0 + c \rightarrow 0$ and $0 + c \rightarrow 1$. (b): The use of a synthetic element to explain the logic behind the observed transformations: ♦$0 + c \rightarrow 0$, ♣$0 + c \rightarrow 1$, $1 + c \rightarrow$ ♦, and $0 + c \rightarrow$ ♣.

For dealing with the combinatorial complexity, at the beginning, the only variables that CALM tries to understand and anticipate are those associated to affective values (what Drescher called primitive values). It is like a built-in reward function that associate positive (desirable) or negative (undesirable) values that the agent can "feel" immediately depending on the value of the variable (i.e. its observation, e.g. true or false if boolean). Staying focused on these variables avoids wasting energy by creating models about everything. Gradually, the variables needed to anticipate the evolution of some important variable (relation of causality) are then considered also important, and the mechanism will seek to model their transformation function too.

Differently from the original mechanism, which creates one initial schema for each action with empty context and result, CALM creates one initial schema for each item on the result vector (which corresponds to a perceptive variable) that has a non neutral affective value, with empty context and action. Each one of these initial schemas will grow up, giving rise to an anticipatory tree, responsible for identifying the relevant elements and the rules that describe the transition of its associated item, allowing in this way to anticipate it. Each anticipatory tree is constructed step by step using 4 operations: *differentiation*, *adjustment*, *integration*, and *abstraction*. Each node in the anticipatory tree is identified by a vector containing observations and actions. Starting from a completely generic root node, where all elements on that vector are set as *undefined*, the agent tries to verify if some regularity persists from the current time-step to the next. The idea is defining a structure that is coherent with the past experience. The expectation of that first schema will be equivalent to the first observation. In this way the schema tries to represent as many transitions as they can be predicted deterministically with the minimum of context (i.e. the minimum of dependencies).

When an observed result contradicts what was expected, the tree is expanded by differentiation, like illustrated in Fig. 6: the contradicted generic schema gives rise to two different children, specializing either one element of context or one element of the action vector, in order to explain the surprise, making the model coherent again. When no primitive action or primitive context element can be used to differentiate a contradicted schema, the mechanism creates a synthetic element of context, representing some abstract or non-observable variable, that, if exists, can explain the incoherence. The schema is then differentiated by this new element. Sometimes, it is not possible to differentiate the schema anymore, in this case, an adjustment is done, setting the contradicted expectation to undefined. Finally, when several schemas are adjusted, some differentiations become unnecessary, then the tree can be pruned, and some leaf schemas integrated in a more general one. In this way, the schemas can be organized in a structure called *anticipatory tree*, which defines a hierarchy, like a decision tree, starting from a completely general context and action at the root node, to most specialized ones, the leaf nodes.

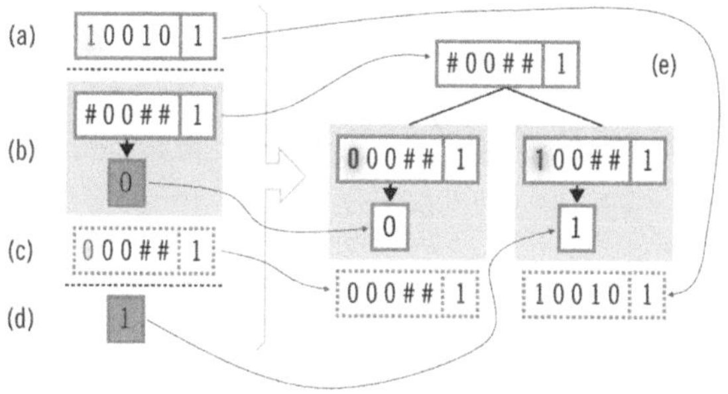

Fig. 6. Differentiation example. (a): An actually experienced situation (with five variables) and executed action (one variable). (b): Activated schema (with compatible context, action, and expectation). (c): Associated episodic memory (representation of actual situations where the scheme has been activated, with no interdependencies between variables). (d): Actually observed result after the execution of the action. (e): Sub-tree generated by differentiation in order to compensate the divergence observed between expectation and result.

To be able to make differentiations, beyond the context and action vectors that identify the schema, the mechanism must store a kind of memory of the effective experienced contexts and actions observed during the past activations of the schema. The identity of the schema corresponding to a maximally generic situation, and the memory corresponding to a maximally specific situation. In order to create a new schema by differentiation, the mechanism searches for an item or action in the memory of the schema that is different from the current situation that is causing the incoherence. The differentiation will preserve the past experience in the old schema, identifying it with a more specialized context or more precise action, separating the new situation in another schema. In terms of computational complexity, the problem is how to manage that memory, because it needs to remember co-occurrences of items and actions. The solution given by Perotto is to explicitly limit the number of variables to observe together, starting from 1, and gradually increasing that number, which in any case will always be low, at the order of few units. The original mechanism proposed by Drescher faces the same problem, and is limited to observe only one additional variable at time.

Figure 7 shows the simple flip problem [45], used to demonstrate CALM's capability to create an abstract variable for representing a relevant non-observable condition of the environment [34]. The set of possible actions is {`l`, `r`, `u`}, standing for *go left*, *go right*, and *unchanged*. The set of possible sensory signals is {`0`, `1`}, corresponding to a single primitive item in Drescher's terms. The difficulty comes from the fact that the agent cannot directly perceive the underlying state of the world. The agent constructed the graph in Fig. 8.

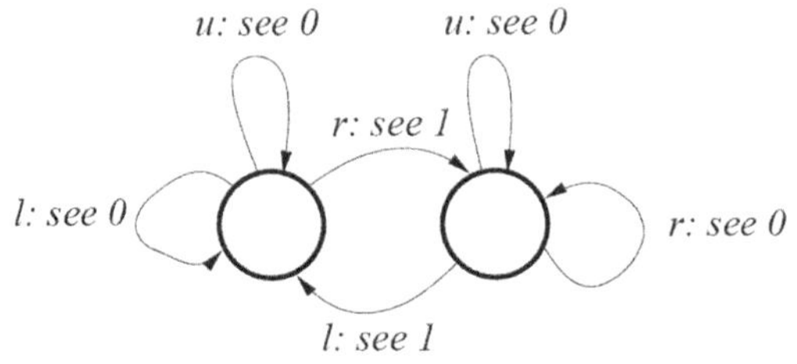

Fig. 7. The flip problem initially proposed by Singh et al. [45]. The `r` action taken from the left states activates the right state and vice versa, yielding sensory signal `1`. All other cases leave the state unchanged and yield sensory signal `0`.

In summary, CALM was used to control an agent that received only a binary sensory signal that does not directly represent the state of the environment. This contrasts with Drescher's experiment in which primitive items represent elements of the state of the world. CALM is able to reconstruct *non observable properties* from sequences of actions and feedback. Similar to Drescher's synthetic items, non observable properties describe the underlying state of the world that cause this feedback. Another addition to Drescher's mechanism is that CALM organizes non observable properties (i.e., synthetic items) hierarchically in the anticipatory trees, which are finally mixed together, as shown in the Fig. 8, which represents the final stable solution for the flip problem.

2.3 Learning Intelligent Decision Agent (LIDA)

The *Learning Intelligent Decision Agent* (LIDA) is a broad cognitive architecture serving both as a conceptual and as a computational framework for modeling the mind [7,8]. A recent version of the LIDA cognitive architecture [24] implements a schema mechanism inspired by Drescher on top of a perceptual memory module called the *Perceptual Associative Memory* (PAM). Percepts are not directly received from the environment but constructed in the PAM through an active sensorimotor process. Other memory modules are implemented including episodic memory, declarative memory, and spatial memory. Knowledge is represented as a directed graph called the *node structure*.

LIDA's schemas work as templates with unbound variables in their context, action, and result. Actions may have unbound variables that qualify and modulate how an action is executed. Schemas are instantiated as *behaviors*.

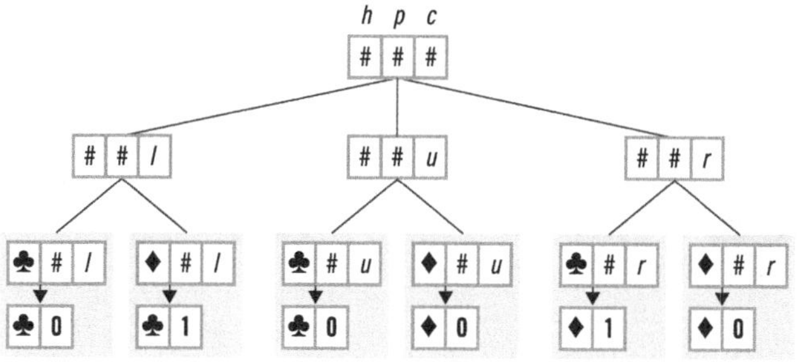

Fig. 8. Final schematic tree for deterministically describing the transition function corresponding to the flip problem [31]. Note that the underlying non-observable left and right states of the problem are represented by ♦ and ♣ values of the abstract synthetic variable. A *Markovian Decision Process* with 4 states (♦0, ♦1, ♣0, ♣1) and a deterministic transition function can be derived from that tree if an immediate reward function is given (e.g. $R(0) = -1, R(1) = +1$), allowing to extract an optimal policy of actions using dynamic programming techniques.

LIDA implements an *affective valence* which drives behavior selection towards desirable outcomes. Liking and disliking are distinguished from wanting and dreading with the argument that they relate to distinct neural pathways in the brain [22]. Wanting and dreading are qualified by *incentive salience* of behaviors.

Kugele [23] recently presented an improved version of LIDA into which the primitive items are replaced by *amodal nodes* organized in the node structure in Perceptual Associative Memory. An important difference between LIDA's amodal nodes and Drescher's primitive items is that amodal nodes may not only represent sensory data but also events. Consequently, LIDA can not only represent a particular context in terms of elements of the world state, but also in terms of sensorimotor events.

Figure 9 shows the cognitive cycle. The dashed rounded rectangle at the bottom shows the schema mechanism as a part of the architecture. The Global Workspace (right) stores a description of the current situation as a graph node. The Procedural Memory module (bottom right) is the memory of schemas. The graph node from Global Workspace activates relevant schemas. The Action Selection module selects an action, primitive or composite, from among the actions of the relevant schemas. Once the action is selected, it is instantiated as a *behavior*, or a *stream of behavior* for composite actions, and their unbound variables are qualified. The behavior is enacted in the Environment (left) and sensory stimuli (signals) are received by the Sensory Memory module (upper left). Sensory signals are processed to produce amodal nodes that represent the current situation. The Attention Codelets module selects some amodal nodes to broadcast in the Global Workspace module.

Kugele [23] demonstrated constructivist procedural learning within LIDA in a multi-armed bandit environment. Different versions of k-armed-bandit envi-

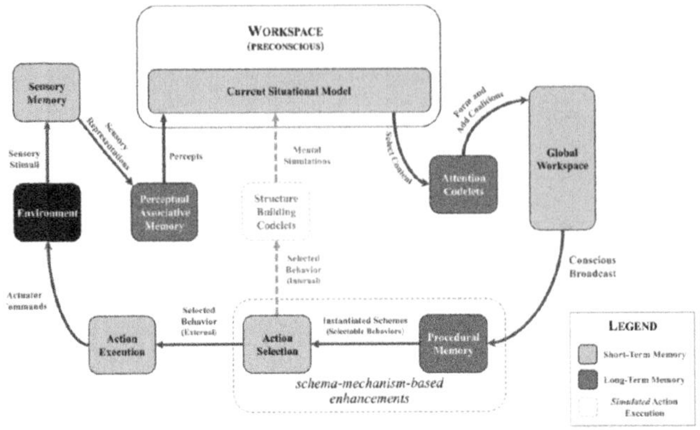

Fig. 9. Simplified depiction of the LIDA cognitive cycle [23, Fig. 4].

ronments have been used previously, for example to demonstrate the theory of *active inference* [46], but Kugele added new difficulties by requiring the agent to sit and to make a deposit before playing. When the agent is standing (State S), it can take the action to sit at a one-armed bandit machine, for example Machine 1 (State M_1). The agent can then pay a deposit (state M_1P), and then play, which may yield a win (State M_1W) with probability $p(W|M_1P)$, or a loss (State M_1L) with probability $p(L|M_1P)$. The agent can choose to stand from any state. Figure 10 represents the state diagram of this environment in the case of two machines.

The set of possible actions for k machines is {`stand`, `sitM1`, ..., `sitMk`, `deposit`, `play`}. The set of initial nodes is {`Win`, `Lose`}, corresponding to a single primitive item in Drescher's terms. Node `Win` is associated with the affective valence (i.e., primitive value) of +1, and Node `Lose` of −1. As in the flip problem presented above, the agent cannot directly perceive the state of the environment. The results obtained in an 8-armed-bandit environment show that after 25,000 interaction cycles, the agent successfully constructed the amodal nodes (i.e., synthetic items) {`S`, `M1`, ..., `M8`, `M1P`, ..., `M8P`} that represent the states of standing or sitting at a machine without or with a paid deposit, along with the schemas representing the transition between states. Once this was learned, the agent determined and chose the machine that yielded the highest payoff.

In summary, LIDA allowed an agent that receives only a binary sensory signal (`Win` or `Lose`) to learn the structure of this environment, and to find adapted behaviors (sitting at the most favorable machine, and then continuing to deposit and to play) satisfying the motivation of maximizing the value won. This is similar to CALM but in a more complex environment.

The agent's adaptation to the k-armed bandit environment was made possible by LIDA's distinction between sensory signal and perceptual memory that allows treating sensory signal as mere events rather than elements of the state. Notably, the agent did not learn the states `MiW` and `MiL` but rather the aggregated state

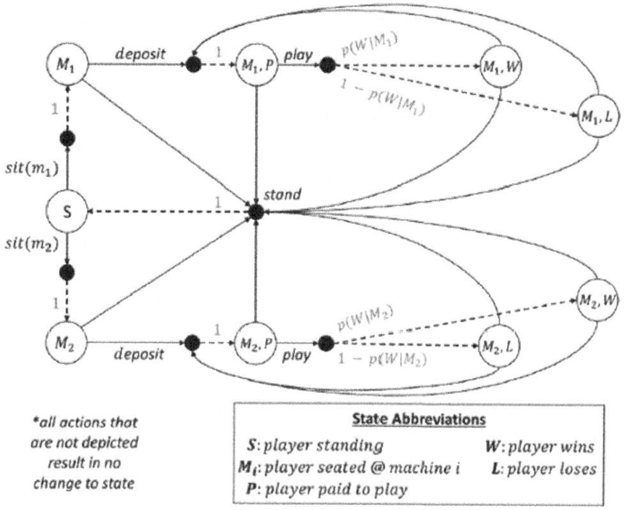

Fig. 10. State diagram of the 2-armed bandit environment [23, Fig. 5]. White nodes depict environmental states. Black nodes depict actions. Links going from white state nodes into black action nodes indicate that this action is taken from that state. Dashed links leaving action nodes are labeled with the probabilities of transitioning to possible states resulting from this action.

`Mi` representing sitting at Machine i. This is because the sensory item (`W` or `L`) is not useful in itself to differentiate significant states from the agent's viewpoint.

Another difference is that LIDA treats composite schemas as procedural knowledge (stream of behavior) that are handled as a whole action by the Action Execution module, which facilitates the learning of the sequence `deposit play`, before finding the machine that yields the highest payoff.

2.4 Enactive Cognitive Architecture (ECA)

Olivier Georgeon and colleagues also studied problems in which the agent receives little sensory information and must find behaviors that satisfy its predefined motivational criteria [16]. Similar to CALM and LIDA, Georgeon's schemas are based on events of interaction that do not necessarily represent elements of relevant states.

Formally, schemas are tuples ⟨pre-subschema, decision, post-subschema⟩ in which the pre-subschema and the post-subschema are other schemas. The mechanism is initialized with a set of *primitive schemas* that define the primitive possibilities of interaction of the agent with its environment. Primitive schemas are initialized with a predefined valence (equivalent to Drescher's primitive value and LIDA's affective valence) that specify the agent's "inborn" preferences of interaction.

In the case of a simulated agent, a primitive schema is the association of a *primitive action*, for example `move-forward` or `turn`, with a *primitive outcome*, for example `bump` or `no-bump`. The primitive schema ⟨`move-forward`, `bump`⟩ initialized with a negative valence specifies an agent that "dislikes" bumping into walls when it tries to move forward.

The mechanism learns new schemas from the bottom up, with higher-level schemas made of a sequence of two previously-learned subschemas, as illustrated in Fig. 11. Schemas are noted with the letter i because they are sometimes called *interactions*. For example, if the primitive schema $i00$, made of action $a0$ and outcome $o0$, was enacted at time $t-2$, and the primitive schema $i11$, made of action $a1$ and outcome $o1$, was enacted at time $t-1$, then the composite schema $\langle i00, a1, i11\rangle$ is learned (added to memory) or its *weight* is incremented if it already existed. The weight of schemas plays a similar role as their reliability in Drescher's implementation.

When a new schema's post-subschema is primitive, the new schema's decision consists of performing the post-subschema's action. When the post-subschema is composite, the new schema's decision consists of enacting the whole sequence specified by the post-subschema. This mechanism constructs a hierarchy of schemas similar to LIDA's hierarchical streams of behaviors.

At time t, the previously learned schemas whose pre-subschemas match the current situation are *activated*. Activated schemas compete to propose their decisions according to a *proclivity value* computed from the activated schema's weight and the post-subschema's valence.

A decision to enact a composite schema generates a stream of behavior consisting of sequentially enacting the whole sequence of primitive schemas that constitute this composite schema. If the enaction of any of its primitive subschemas fails (i.e., the actual outcome differs from the schema's outcome), then the enaction of the sequence is interrupted.

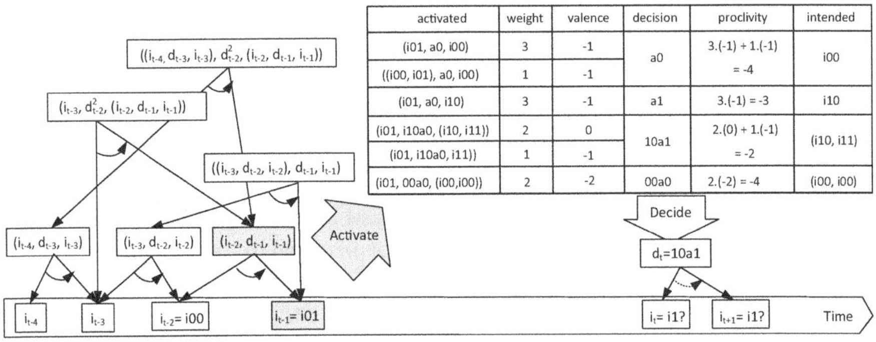

activated	weight	valence	decision	proclivity	intended
(i01, a0, i00)	3	-1	a0	3.(-1) + 1.(-1) = -4	i00
((i00, i01), a0, i00)	1	-1			
(i01, a0, i10)	3	-1	a1	3.(-1) = -3	i10
(i01, i10a0, (i10, i11))	2	0	10a1	2.(0) + 1.(-1) = -2	(i10, i11)
(i01, i10a0, i11))	1	-1			
(i01, 00a0, (i00,i00))	2	-2	00a0	2.(-2) = -4	(i00, i00)

Fig. 11. Schema learning and selection. Schemas are nested tuples: ⟨pre-subschema, decision, post-subschema⟩. Over time, new decisions and new schemas are learned from the bottom up. Recently enacted schemas (in gray) activate the previously-learned higher-level schema whose pre-subschema they match. Activated schemas propose their post-subschemas with a proclivity value calculated from the activation weight and the expected valence. The schema with the highest proclivity is selected to try to enact.

Georgeon [12] reports an experiment in the environment shown in Fig. 12 and shared in a video [10]. The set of primitive actions is {`move-forward`, `turn-left`, `turn-right`, `feel-front`, `feel-left`, `feel-right`}. The set of possible outcomes is {`no-wall`, `wall`} corresponding to a single primitive item. The schema ⟨`move-forward`, `no-wall`⟩ has a valence of +5 making it the only "enjoyable" interaction to the agent. The schema ⟨`move-forward`, `wall`⟩, corresponding to bumping into wall, has a valence of −10 making it strongly dreaded. The 6 schemas that have a `feel` action have a valence of −1 making them slightly costly. The schemas that have a `turn` action have a valence of −3 making them mildly costly.

Like in the experiments with CALM and LIDA presented above, the difficulty comes from the fact that the agent only has a single binary sensory signal and cannot directly sense the state of the environment. The agent thus characterizes the state of the environment through enacted schemas. Results show that after approximately a hundred interaction cycles, the agent learns to use the `feel` schemas to actively sense the environment. This active sensing provides a situation awareness that makes the agent turn to empty cells when necessary, avoid bumping into walls, and optimize its moving forward.

Fig. 12. The small loop Challenge [12]. The agent can move forward to an empty (white) cell, bump into a wall (green cell), turn to the left or to the right by 90°, or "feel" the cell in front, to the left, or to the right. The sensory signal is a single-bit feedback from the action ("empty" or "wall"). The agent progressively learns to used the "feel" schemas to sense its surrounding environment, not move forward when it feels a wall in front, and turn to a direction where it feels no wall, thus maximizing the "move forward" interaction that has a predefined positive valence.

In a robot, primitive actions are defined as hard-coded control loops that regulate actuators and monitor sensory feedback over a predetermined period. For example, the `move-forward` action turns on the wheel engines and runs a loop that monitors the elapsed time and the collision detection sensor. If a collision is detected, the schema is terminated before its predetermined duration. Upon termination, the wheel engines are turned off and the outcome is sent to the schema mechanism, for example `no-bump` or `bump`.

To deal with the complexity of the physical world, Georgeon and colleagues encapsulated the schema mechanism in a cognitive architecture called Enactive Cognitive Architecture (ECA) [14]. ECA incorporates a spatial memory module that can simulate the enaction of schemas to help select the most desirable schema in the current spatio-temporal context. Like in LIDA, schemas may include unbound variables that can be qualified upon their instantiation. For example, the `turn` action has the variable `angle` that specifies the rotation span in the horizontal plane (yaw). When the schema mechanism module selects the `turn` action, the spatial memory module qualifies the angle to the desired direction.

ECA was demonstrated with the robot shown in Fig. 13 [11]. When the robot finds reliable schemas in a consistent spatial localization, it creates a datastructure called a *phenomenon* that represents the "thing" in the world that affords these schema. This is analogous to the construction of synthetic items in Drescher's schema mechanism.

Fig. 13. Robot learning a phenomenon (synthetic item) representing a cylindrical object [17, Fig. 5]. Left: Photo of the robot showing its ultrasonic echo-localization sensor mounted on its pivoting head. Right: Representation of the *cylindrical phenomenon.* The phenomenon is aggregated from many sensorimotor "manifestations" of the cylindrical object through echo-localization measures. Blue triangles represent the cones of the ultrasonic signals sent by the robot over time from different positions around the moneybox. The black line is the shape of the cylindrical object reconstructed from the echoes.

In summary, ECA's schema mechanism uses schemas as atomic elements for constructing higher-level schemas instead of sensory signals, synthetic items, and actions. This design choice facilitates the hierarchical leaning of composite schemas because they are not based on abstract representations of elements in the world. In the experiment reported in Fig. 12, the agent manages to adapt to its environment using only schemas but no synthetic items. Like Drescher and Kugele, Georgeon implemented an effect of *habituation* that causes reliable schemas to be preferably reused, which facilitates the construction of higher-level schemas on top of them. This habituation principles has echoes in Woolford's *sensorimotor sequence reiterator* [52].

ECA makes a strong design choice by assuming that the environment has a three-dimensional Euclidean structure and is populated with physical objects. This assumption is coded in the cognitive architecture through the spatial memory module and by adding unbound variables to schema that encode spatial parameters. This design choice allows ECA to construct phenomena to represent "things" in the physical world. We hypothesize that it can help avoid the scalability limitation of Drescher's implementation of synthetic items. Including spatial variable in schemas and qualifying them adequately, however, raises many questions that remain open.

2.5 Autocatalytic Endogenous Reflective Architecture (AERA)

AERA's models [27,49,50], similar to Drescher's schemas [5] and the neuro-symbolic approach presented in [21], represent knowledge by contextualized programs that capture antecedental and successional states of important relationships (e.g. the relationship between a hand and an object that one wants to pick up). Unlike these, AERA places explicit representation of cause-effect relations at its core [51]. During its continuous learning, driven by a self-maintained dynamic goal hierarchy, the most useful (effective and efficient) models of cause-effect relations encountered in the environment are retained, while others are discarded. Based on predictions, informed experimentation, and strategic induction, the system autonomously creates new knowledge cumulatively, effortlessly unifying new information with old, even in light of contradictions. Further separating AERA from other approaches are its atomic elements for knowledge creation, whose operational semantics are transparent to the system's own reasoning, meaning that they can be inspected and learned by the same explicit, defeasible unified reasoning mechanisms as used for learning about the world.

Key representational components of this approach include [41,43]:

- **Facts** are statements that represent a cognitive event, e.g. operations happening in AERA itself; facts are the only entity in AERA that have complete certainty (corresponding to Descartes' axiom "I think, therefore I am").
- A **composite State model (CST)** captures a set of simultaneous (timeless) Facts that the AERA system considers to be (potentially) useful and/or necessary for properly predicting an event's consequences; used to e.g. represent a variety of subsets in the world such as spatial relations of objects, the values of a perceived object's properties, such as its position and color, etc.
- A **cognitive event model (CEM)** captures a (hypothesized) causal influence of a Fact on another Fact; used to represent the effect that an action or event in particular circumstances, e.g. a command initiated by AERA at some timestep, has on the state of the world at a later timestep. When a CEM fires it produces a prediction (while the production of a prediction is represented as a Fact, the confidence in the prediction itself is defeasible and is in part based on the confidence of the CEM).

- A **requirement model** (M_{req}) captures the (predicted) requirements/conditions that must be met for a CEM to be relevant in a particular context at a particular time.

AERA's programming paradigm allows the system itself to inspect these types of information structures, use them "for parts" to create variations, as well as create new ones from scratch. AERA's programming language, Replicode, allows it to autonomously use these building blocks for self-programming, in the form of coherent plans, predictions, and goals that it can pursue [25,26,48].

Both *CEM*s and M_{req}s have a left-hand side (LHS) and a right-hand side (RHS), composed of sets of Facts and constraints (e.g., math equations). The LHS Facts describe (assumed) conditions under which the RHS Facts would be produced. Figure 14 shows how the above components lead to making a prediction. When some sensory Facts match everything in a *CEM*'s LHS, a relevant M_{req} allows the *CTS*'s instantiation, which means the conditions for making a prediction via a *CEM* are met. A successful instantiation of a *CEM* means that everything in its RHS will happen. As the *CEM* predicts a future state after an event's occurrence, this is verified and marked as a "successful prediction" by AERA only if the prediction matches the state that is observed.

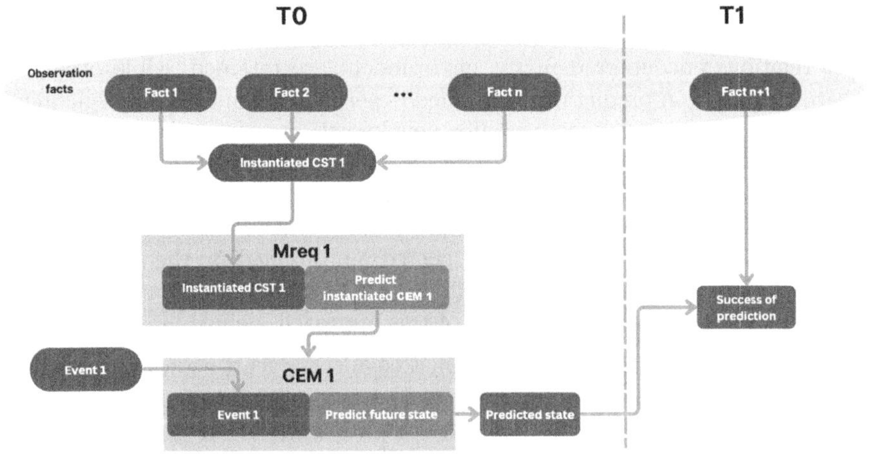

Fig. 14. The connection between CST, M_{req}, and CEM. Time T0 (left): The observation facts matching $CST1$ allow M_{req} to trigger a prediction by instantiating $CEM1$. Time T1 (right): AERA checks whether the observations match the predicted state.

As an example, the triad of $CST1$, $M_{req}1$, and $CEM1$ can represent the physical movement behavior of an entity when it collides with another entity (e.g., a hand). $CEM1$ represents the dynamics of the movement, connecting the collision event (left-hand side of $CEM1$) to the moved entity's new position (right-hand side of $CEM1$). $CST1$ represents the preconditions of the behavior, such as the entities' position, weight, etc., and the fact that the point of collision and the moved entities' position must initially be the same. $M_{req}1$ predicts that if a striking entity hits another entity with the conditions described by $CST1$, the stricken entity's position will change a specific amount in the next time step.

The interactions of the operational principles of AERA are quite intricate in their entirety; important aspects that are covered elsewhere include induction - the creation of new knowledge - unified abduction-deduction - the creation of plans and explanations - and runtime relevance computations [26]. The unified integration of these operational principles in a single coherent architecture is what gives AERA its power, allowing its seed - initial human-provided knowledge - to be orders of magnitude smaller than what AERA can create on its own through learning from experience.

These mechanisms have been demonstrated for control of virtual robots, learning and using language, and multimodal interaction [50]. Figure 15 shows the latest experiment reported in [41] and shared in video [42]. The task has a *guided experimentation* phase, during which the experimenter tele-operates the robot, and a *task solving* phase, during which the robot must autonomously learn to grasp new objects it has never grasped before, which are novel in shape but familiar in size.

During the guided phase, commands and successful grasping events makes AERA learn triads of Causal Event Models (CEMs), Composite State models (CSTs), and requirement models (Mreqs) for each command. During the task-solving phase, the analogy process induces new grasping models based on the objects' familiarity. Results show that, after several trials unsuccessfully testing the learned models, AERA induces new CSTs and Mreqs capable of triggering the appropriate actions.

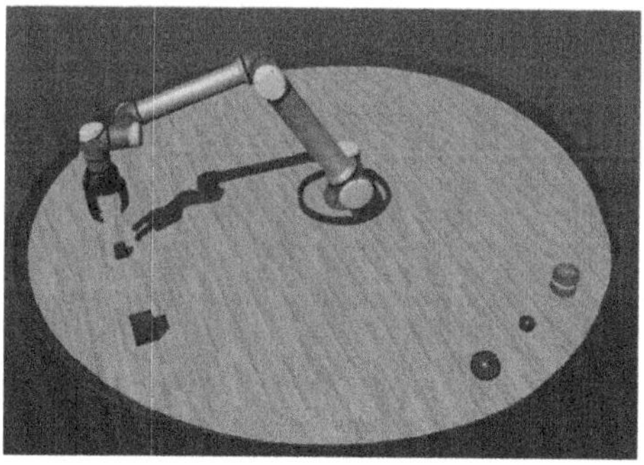

Fig. 15. Experimentation from [44, Fig. 2]. In the guided phase, the robot is tele-operated to grab the small and big cubes on the left; in the task solving phase, it quickly learns to autonomously grab the small ball, big ball, and big cylinder on the right by analogy.

In summary, AERA introduces the capacity for the schema mechanism to process events that have en external cause in addition to those caused by the actions initiated by the agent. This is done by distinguishing the context in which the event occurs, represented by Composite State models (CSTs), from the hypothetical causal effects of the event stored in Cognitive Event Models (CEMs). The agent can thus construct hypothetical causal models as it passively observes other agents interacting with the world or as it observes its own interactions while being tele-operated. The agent can then test the hypothetical causal models by itself to evaluate their reliability.

Both the events caused by the agent's actions and those passively observed generate the instantiation of cognitive events and the generation of *prediction facts*. AERA considers both external events and cognitive events as actual facts that AERA can internally process, which opens the way to reflexivity.

3 Conclusion

Schema mechanisms are characterized by their dynamic approach to constructing concepts from experience of interaction. Compared to more classical techniques such as statistical symbol grounding [20] or active inference [9], this dynamic approach takes into account counterfactual sequences of interactions for the construction of concepts. This makes schema mechanisms particularly reflect Piaget's core idea that sensorimotor schemas serve as the initial instruments for constructing knowledge.

The four new schema mechanisms presented here, CALM, LIDA, ECA, and AERA, bring an additional idea to the picture: the capacity to deal with sensory signals (primitive items) that are not "percepts representing elements of the world", but mere feedback from action or events of interaction. In so doing, these four new schema mechanisms address a limitation pointed out by Bettoni when he wrote that "Drescher's constructivism is not Piaget's constructivism, mainly because of its tacit acceptance of cognitive dogmatism" [2, p. 6]. We interpret this critique as suggesting that using primitive items that represent elements of the state falls within cognitive dogmatism because it implements the assumption that the agent has direct access to the ontology of the world. In contrast, constructing schemas from interactional events better reflects Piaget's idea that knowledge arises from an initial "complete un-differentiation between the subject and the object".

Basing schemas on interactional events is particularly important when designing the agent's control system. Drescher's agents sought goal states defined in the environment—a form of *extrinsic motivation* that is inherently prone to scalability issues as the complexity of the environment grows. In contrast, the new schema mechanisms seek rewarding interactions as illustrated in the experiments above: CALM: moving left and right, LIDA: winning, ECA: moving forward, and AERA: grabbing objects. This is a form of *interactional motivation* [13] whose complexity may not grow as the complexity of the environment grows, allowing easier transfer to robots in the open world.

LIDA and ECA have demonstrated interactional motivation using both negatively valenced primitive schemas (e.g., LIDA: `deposit`, ECA: `feel` and `turn`) and positively valenced ones (LIDA: `win`, ECA: `move-forward`). They learn composite actions made of sequences of both kinds of schemas (LIDA: ⟨`deposit`, `win`⟩, ECA: ⟨`turn`, `move-forward`⟩), which allows the emergence of behaviors adapted to delayed reward. These sequences work as short programs that the agent can re-execute, which leads to self-programming—an important aspect of constructivist epistemology [15].

CALM has focused on the construction of hierarchies of synthetic items, exploring Piaget's idea of assimilation and accommodation. LIDA, ECA, and AERA have encapsulated the schema mechanism within a cognitive architecture. They handle schemas with unbound variables, reducing the number of schemas and helping manage complexity, albeit at the cost of potentially diminishing generality. These cognitive architectures internally simulate the enaction of the schemas before selection and qualification. AERA has further explored this direction by allowing reflexivity upon internal events. Altogether, the four new schema mechanisms extend our vision of how Piaget's ideas can be translated into artificial intelligence. Because they incorporate significantly more of constructivist epistemology than Drescher's original version, we propose to call them schema mechanisms 2.0. Figure 16 synthesizes their contributions.

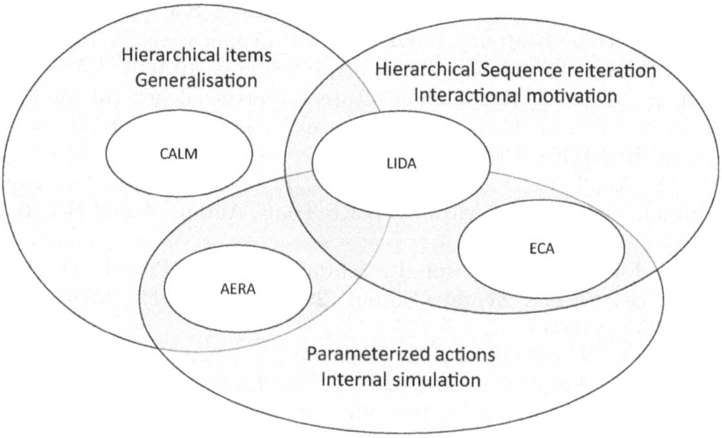

Fig. 16. Coverage of principal dimensions by schema mechanisms 2.0.

Schema Mechanisms 2.0 have still not passed the sensorimotor stage of development—the first defined by Piaget. The question whether their implementation in computers is overly reductionist compared to biological systems remains open. They nonetheless demonstrate some promising results with robots. LIDA and ECA have implemented an Euclidean spatial memory module and an intuitive physics engine to handle schemas that encode physical properties. This is a strong commitment that reduces generality but may open the way to developmental artificial intelligence for autonomous robots in the open world.

Acknowledgments. This article benefited from insightful discussions with Gary Drescher and Jeff Thompson.

Disclosure of Interests. The authors have no competing interests to declare that are relevant to the content of this article.

References

1. Babakr, Z.H., Mohamedamin, P., Kakamad, K.: Piaget's cognitive developmental theory: critical review. Educ. Q. Rev. **2**(3) (2019). https://doi.org/10.31014/aior.1993.02.03.84
2. Bettoni, M.C.: Made-up minds: a constructivist approach to artificial intelligence - a book review. AI Commun. **6**(3), 234–240 (1993). https://doi.org/10.3233/AIC-1993-63-413
3. Drescher, G.L.: A mechanism for early piagetian learning. In: Forbus, K.D., Shrobe, H.E. (eds.) Proceedings of the 6th National Conference on Artificial Intelligence, Seattle, WA, USA, pp. 290–294. Morgan Kaufmann (1987)
4. Drescher, G.L.: Made-up minds: a constructivist approach to artificial intelligence. Ph.D. thesis, MIT (1989)
5. Drescher, G.L.: Made-up minds: a constructivist approach to artificial intelligence. Artificial Intelligence. MIT Press (1991)
6. Drescher, G.L.: The schema mechanism. In: Hanson, S.J., Remmele, W., Rivest, R.L. (eds.) Machine Learning: From Theory to Applications. LNCS, vol. 661, pp. 125–138. Springer, Heidelberg (1993). https://doi.org/10.1007/3-540-56483-7_27
7. Franklin, S.: A foundational architecture for artificial general intelligence. In: Advances in Artificial General Intelligence: Concepts, Architectures and Algorithms, pp. 36–54. IOS Press (2007)
8. Franklin, S., Madl, T., D'mello, S., Snaider, J.: Lida: a systems-level architecture for cognition, emotion, and learning. IEEE Trans. Auton. Mental Dev. **6**(1), 19–41 (2013)
9. Friston, K., FitzGerald, T., Rigoli, F., Schwartenbeck, P., Pezzulo, G.: Active inference: a process theory. Neural Comput. **29**(1), 1–49 (2017). https://doi.org/10.1162/NECO_a_00912
10. Georgeon, O.L.: The small loop challenge (2012). https://youtu.be/Mstl8BRCbSI
11. Georgeon, O.L., Lurie, D., Robertson, P.: Artificial enactive inference in three-dimensional world. Cogn. Syst. Res. **86**, 101234 (2024). https://doi.org/10.1016/j.cogsys.2024.101234
12. Georgeon, O.L., Marshall, J.: The small loop problem: a challenge for artificial emergent cognition. In: International Conference on Biologically Inspired Cognitive Architectures, pp. 137–144 (2012)
13. Georgeon, O.L., Marshall, J.B., Gay, S.: Interactional motivation in artificial systems: between extrinsic and intrinsic motivation. In: 2012 IEEE International Conference on Development and Learning and Epigenetic Robotics (ICDL), pp. 1–2. IEEE (2012). https://doi.org/10.1109/DevLrn.2012.6400833
14. Georgeon, O.L., Marshall, J.B., Manzotti, R.: ECA: an enactivist cognitive architecture based on sensorimotor modeling. Biol. Inspired Cogn. Archit. **6**, 46–57 (2013). https://doi.org/10.1016/j.bica.2013.05.006
15. Georgeon, O.L., Riegler, A.: CASH only: constitutive autonomy through motorsensory self-programming. Cogn. Syst. Res. **58**, 366–374 (2019). https://doi.org/10.1016/j.cogsys.2019.08.006

16. Georgeon, O.L., Ritter, F.E.: An intrinsically-motivated schema mechanism to model and simulate emergent cognition. Cogn. Syst. Res. **15–16**, 73–92 (2012). https://doi.org/10.1016/j.cogsys.2011.07.003
17. Georgeon, O.L., Vidal, J.R., Knockaert, T., Robertson, P.: Simultaneous localization and active phenomenon inference (SLAPI). In: Thórisson, K.R. (ed.) Proceedings of the Third International Workshop on Self-Supervised Learning. Proceedings of Machine Learning Research, vol. 192, pp. 77–88. PMLR (2022)
18. Glasersfeld, E.v.: Radical constructivism: a way of knowing and learning. Studies in Mathematics Education Series, No. 6. Falmer Press (reprinted edn)
19. Guerin, F., Kruger, N., Kraft, D.: A survey of the ontogeny of tool use: from sensorimotor experience to planning. IEEE Trans. Auton. Ment. Dev. **5**(1), 18–45 (2013). https://doi.org/10.1109/TAMD.2012.2209879
20. Harnad, S.: The symbol grounding problem. Physica D **42**(1), 335–346 (1990). https://doi.org/10.1016/0167-2789(90)90087-6
21. Komrusch, S., Minsky, H.: Symbolic guidance for constructivist learning by neural model. In: Proceedings of the Third International Workshop on Self-Supervised Learning, pp. 63–76. PMLR. https://proceedings.mlr.press/v192/komrusch22a.html. ISSN: 2640-3498
22. Kringelbach, M., Berridge, K.: The functional neuroanatomy of pleasure and happiness. Discov. Med. **9**, 579–87 (2010)
23. Kugele, S.: Constructivist procedural learning for grounded cognitive agents. Cogn. Syst. Res. **90**, 101321 (2025). https://doi.org/10.1016/j.cogsys.2025.101321
24. Kugele, S., Franklin, S.: Learning in LIDA. Cogn. Syst. Res. **66**, 176–200 (2021). https://doi.org/10.1016/j.cogsys.2020.11.001
25. Nivel, E., Thórisson, K.R.: Replicode: a constructivist programming paradigm and language. Technical RUTR-SCS13001, Reykjavik University School of Computer Science (2013)
26. Nivel, E., Thórisson, K.R.: Towards a programming paradigm for control systems with high levels of existential autonomy. In: International Conference on Artificial General Intelligence, pp. 78–87. Springer (2013)
27. Nivel, E., et al.: Autocatalytic endogenous reflective architecture (2013)
28. Oudeyer, P.Y., Kaplan, F., Hafner, V.V.: Intrinsic motivation systems for autonomous mental development. IEEE Trans. Evol. Comput. **11**(2), 265–286 (2007). https://doi.org/10.1109/TEVC.2006.890271
29. Perotto, F.S.: Um Mecanismo Construtivista para Aprendizagem de Antecipações em Agentes Artificiais Situados. Ph.D. thesis, Universidade Federal do rio Grande do Sul (UFRGS), Porto Alegre, Brazil (2010)
30. Perotto, F.S.: Un Mécanisme Constructiviste d'Apprentissage Automatique d'Anticipations pour des Agents Artificiels Situés. Ph.D. thesis, National Polytechnic Institute of Toulouse (INP), France (2010). http://www.theses.fr/2010INPT0041/document. Thèse de doctorat dirigée par Buisson, Jean-Christophe et Campos Álvares, Luís Otávio Intelligence artificielle Toulouse, INPT 2010
31. Perotto, F.S.: A computational constructivist model as an anticipatory learning mechanism for coupled agent environment systems. Constr. Found. **1**(9), 46–56 (2013)
32. Perotto, F.S., Alvares, L.O.: Incremental inductive learning in a constructivist agent. In: Bramer, M., Coenen, F., Tuson, A. (eds.) Research and Development in Intelligent Systems XXIII, Proceedings of AI-2006, the Twenty-Sixth SGAI International Conference on Innovative Techniques and Applications of Artificial Intelligence, Cambridge, UK, 11–13 December 2006, pp. 129–142. Springer (2006). https://doi.org/10.1007/978-1-84628-663-6_10

33. Perotto, F.S., Alvares, L.O.: Learning regularities with a constructivist agent. In: Nakashima, H., Wellman, M.P., Weiss, G., Stone, P. (eds.) 5th International Joint Conference on Autonomous Agents and Multiagent Systems (AAMAS 2006), Hakodate, Japan, 8–12 May 2006, pp. 807–809. ACM (2006). https://doi.org/10.1145/1160633.1160778
34. Perotto, F.S., Buisson, J.C., Alvares, L.O.C.: Constructivist anticipatory learning mechanism (calm): dealing with partially deterministic and partially observable environments. In: International Conference on Epigenetic Robotics (EpiRob), Piscataway, NJ, USA, pp. 117–127. Lund University Cognitive Science (2007)
35. Piaget, J., Inhelder, B.: The Psychology of the Child, 2 edn. Basic Books (1969)
36. Piaget, J.: The Construction of Reality in the Child. Routledge and Kegan Paul (1955)
37. Piaget, J.: The Origin of Intelligence in the Child. Routledge and Kegan Paul (1955)
38. Piaget, J.: The Principles of Genetic Epistemology. Psychology Press (1997)
39. Piaget, J.: L'épistémologie génétique. Quadrige, PUF (2011)
40. Raju, R.V., Guntupalli, J.S., Zhou, G., Lázaro-Gredilla, M., George, D.: Space is a latent sequence: structured sequence learning as a unified theory of representation in the hippocampus arXiv:2212.01508 (2022). http://arxiv.org/abs/2212.01508
41. Sheikhlar, A.: Autonomous causal generalization (2024)
42. Sheikhlar, A.: Sizemattersgrab experiment (2024). https://youtu.be/JXgdSjU-7OI
43. Sheikhlar, A., Thórisson, K.R.: Causal generalization via goal-driven analogy. In: International Conference on Artificial General Intelligence, pp. 165–175. Springer (2024)
44. Sheikhlar, A., Thórisson, K.R.: Causal generalization via goal-driven analogy. In: Thórisson, K.R., Isaev, P., Sheikhlar, A. (eds.) Artificial General Intelligence, vol. 14951, pp. 165–175. Springer (2024). https://doi.org/10.1007/978-3-031-65572-2_18
45. Singh, S., Littman, M.L., Jong, N.K., Pardoe, D., Stone, P.: Learning predictive state representations. In: Proceedings of the Twentieth International Conference on International Conference on Machine Learning, ICML 2003, pp. 712–719. AAAI Press (2003)
46. Smith, R., Friston, K.J., Whyte, C.J.: A step-by-step tutorial on active inference and its application to empirical data. J. Math. Psychol. **107**, 102632 (2022). https://doi.org/10.1016/j.jmp.2021.102632
47. Sutton, R.S., Barto, A.G.: Reinforcement Learning: An Introduction, 2 edn. The MIT Press (2018)
48. Thórisson, K.R.: A new constructivist AI: from manual methods to self-constructive systems. In: Theoretical Foundations of Artificial General Intelligence, pp. 145–171. Springer (2012)
49. Thórisson, K.R., Bieger, J., Li, X., Wang, P.: Cumulative learning. In: Hammer, P., Agrawal, P., Goertzel, B., Iklé, M. (eds.) AGI 2019. LNCS (LNAI), vol. 11654, pp. 198–208. Springer, Cham (2019). https://doi.org/10.1007/978-3-030-27005-6_20
50. Thórisson, K.R., et al.: Autonomous acquisition of natural situated communication. IADIS Int. J. Comput. Sci. Inf. Syst. **9**(2), 115–131 (2014)
51. Thórisson, K.R., Talbot, A.: Cumulative learning with causal-relational models. In: Iklé, M., Franz, A., Rzepka, R., Goertzel, B. (eds.) AGI 2018. LNCS (LNAI), vol. 10999, pp. 227–237. Springer, Cham (2018). https://doi.org/10.1007/978-3-319-97676-1_22

52. Woolford, F.M.G., Egbert, M.D.: A precarious sensorimotor sequence reiterator for modelling enactive habits. In: ALIFE 2020: The 2020 Conference on Artificial Life, pp. 771–779. MIT Press (2020). https://doi.org/10.1162/isal_a_00249
53. Ziemke, T.: The construction of 'reality' in the robot: constructivist perspectives on situated artificial intelligence and adaptive robotics. Found. Sci. **6**(1), 163–233 (2001). https://doi.org/10.1023/A:1011394317088

Intrinsic Motivation for Artificial Agents

Paul Robertson[1(✉)] and Olivier L. Georgeon[2]

[1] Dynamic Object Language Labs, Lexington, MA 02421, USA
paulr@dollabs.com
[2] UR CONFLUENCE: Sciences et Humanites (EA 1598), UCLy, Lyon, France
ogeorgeon@univ-catholyon.fr

Abstract. The purpose of this paper is to review the state-of-the-art in intrinsic motivation, specifically in the absence of explicit goals. This paper focuses on how intrinsic motivation is used in constructivist machine learning to foster exploration, learning, and adaptability in complex environments. In particular, this paper provides a survey of several kinds of intrinsic motivation and discusses how they can be used.

1 Introduction

Why do humans learn, explore, and create without external incentives? The answer lies in intrinsic motivation: a drive to engage in activities for their own sake, fueled by curiosity, mastery, autonomy, or inherent satisfaction, rather than external rewards. Intrinsic motivation is fundamental to human learning, creativity, and participation, enabling individuals to adapt, grow, and thrive. But can machines, particularly AI agents and robots, be endowed with a similar capacity? This paper explores how intrinsic motivation can inspire the development of autonomous, adaptable AI agents, especially in environments devoid of explicit goals.

Artificial intelligence has made great strides in enabling agents to perform designated tasks efficiently. However, a critical challenge remains: How should an AI agent decide whether to perform its assigned duties or engage in learning and exploration? Human learning provides clues. People learn in various ways, either through being taught, mimicking others, exploring their environment, or even through play. This continuous context-driven process allows humans to adapt and acquire new skills over time. AI agents, too, must develop similar capabilities to navigate and respond to complex, dynamic environments effectively.

Current robotic systems often rely on explicit instructions or teleoperation to perform their tasks. Although practical, these methods impose significant labor and cognitive burdens on humans. Explicitly programmed goals offer some relief, but still require considerable human oversight and programming effort. For robots to truly integrate into human environments and achieve the autonomy needed to reduce human workload, they must not only execute tasks, but also learn and adapt as humans do: developing new skills and refining existing ones.

P. Robertson and O. Georgeon (Eds.): IWSSL 2024, LNCS 15457, pp. 88–120, 2025.
https://doi.org/10.1007/978-3-031-96325-4_5

To bridge this gap, intrinsic motivation mechanisms hold promise. By mimicking the human drive to explore and learn for inherent satisfaction, these mechanisms can empower AI agents to autonomously engage in meaningful behaviors without constant reliance on extrinsic rewards or predefined objectives. Such agents could choose actions based on curiosity, the desire to master skills, or the potential to discover novel solutions. This self-directed behavior aligns closely with the vision of creating adaptable and collaborative robots capable of seamlessly integrating into human environments.

Understanding intrinsic motivation in humans provides a valuable foundation for designing similar mechanisms in AI. Human behavior often lies at the intersection of necessity and inherent satisfaction. Eating, for example, serves both survival needs and social or sensory pleasures. Similarly, work can be a source of inherent satisfaction for some and a mere necessity for others. These blurred boundaries offer insight into how intrinsic and extrinsic motivations coexist and drive complex behaviors.

The key lies in replicating the psychological drives that underpin human behavior. Self-Determination Theory (SDT) [9,10,27,28,35] provides a robust framework to understand how autonomy, competence, and relatedness shape human motivation. Using these insights, we can design AI agents that not only perform tasks efficiently but also learn, explore, and grow autonomously, much like humans do. Such agents would not only be more functional but also more relatable and well-received by their human counterparts.

This paper reviews the state-of-the-art in intrinsic motivation research, focusing on its application in the absence of explicit goals. By drawing parallels to human motivation and exploring its potential in AI, we aim to illuminate a path toward creating more autonomous, adaptable, and human-aligned artificial agents.

2 Self Determination Theory

We begin by looking at SDT to see how inherent satisfaction relates to intrinsic motivation. SDT enumerates a number of sources of inherent satisfaction in humans as follows.

Autonomy. The sense of autonomy empowers individuals to exercise control over their actions and make choices freely. This includes pursuing hobbies or interests without external pressure and working on projects that align with personal values. Autonomy fosters a sense of ownership and personal agency, which is intrinsically rewarding.

Competence. The acquisition of proficiency in a skill, the attainment of objectives or the successful resolution of obstacles, such as solving complex problems, completing demanding tasks, or improving performance in a sport, musical instrument, or craft, all contribute to the development of a sense of self-efficacy and personal growth.

Relatedness. Human beings are inherently social creatures, and a sense of connection to others is essential for our well-being. This connection can manifest itself in various forms, such as belonging to a community, forming meaningful relationships, and collaborating on shared projects with a team. Engaging in deep conversations with friends and loved ones also contributes to our emotional security and sense of belonging.

Curiosity and Discovery. The exhilaration of venturing into new concepts, acquiring fresh knowledge or uncovering concealed patterns, whether through reading about an intriguing topic or conducting research, exploring a new destination or trying an unfamiliar activity. Curiosity taps into our inherent propensity to learn and comprehend the world.

Flow State. A state of profound immersion where an individual becomes fully absorbed in an activity, losing track of time. This state can be achieved through activities such as writing, painting, or coding for extended periods without conscious awareness of the elapsed time. Similarly, intense focus on playing a musical instrument or participating in sports can lead to a state of deep engagement and fulfillment. The harmonious balance between challenge and skill contributes to a profound sense of contentment and satisfaction.

Creativity and Self-expression. The creative process involves the generation of new and distinctive ideas or expressions, manifesting in various forms such as poetry, visual arts, music composition and problem solving. Creativity empowers individuals to authentically express their thoughts, emotions, and identities, enabling them to bring their unique visions to fruition.

Purpose and Meaning. Engaging in activities that align with one's values or contribute to a greater purpose, such as volunteering for a cause that deeply resonates with one's beliefs or working on a project that positively impacts one's community, can significantly enhance life's meaning and foster emotional well-being.

Mastery of Challenges. Overcoming obstacles that are neither too easy nor too difficult, such as completing a challenging puzzle or ascending a difficult trail, or reaching a significant milestone after consistent effort, fosters resilience and personal accomplishment.

Playfulness and Joy. Engaging in activities solely for the purpose of enjoyment, amusement, or recreation, such as playing games, engaging in humorous banter, dancing, singing, or pursuing spontaneous endeavors. Play serves as a valuable stress reliever, fosters creativity, and contributes to overall well-being and happiness.

Contribution and Impact. The satisfaction derived from helping others or leaving a positive impression, such as mentoring someone, imparting knowledge, or building something that benefits a community, is profound. Contributing to the well-being of others fosters a profound sense of purpose and gratitude.

Alignment with Personal Values. Acting according to deeply held beliefs and principles, such as advocating for justice and fairness or practicing kindness, honesty and generosity. Living authentically reinforces integrity and self-respect.

Physical and Sensory Enjoyment. Deriving pleasure from bodily sensations and physical activities, including savoring a delicious meal, indulging in a warm bath, experiencing exhilaration during physical exercise or nature walks. Sensory experiences are directly associated with the immediate experience of pleasure and relaxation.

Not surprisingly, many of the inherent satisfactions listed above are covered by intrinsic motivation implementations that are enumerated in Sect. 4.

It can be argued that all the inherent satisfactions mentioned above are present in all humans to varying degrees, implying that some of these inherent satisfactions exert themselves more strongly than others. No single measure of intrinsic motivation is sufficient; rather, we aim for all measures to be present and for there to be some mechanism that decides the strength of each motivational dimension or inherent satisfaction. For example, for each inherent motivation, a coefficient representing the strength of the corresponding satisfaction could be used to determine to what extent each intrinsic motivation influences behavior. In this paper we wish to reinforce the view that if all such motivations exist in humans, we should expect the same in a general A.I. agent, although for a less general A.I. agent, a limited number of intrinsic motivations might be adequate. In Sect. 6 we describe an algorithm that integrates multiple motivations in order to generate behavior that drives a simple robot to learn.

Inherent satisfaction refers to actions that provide personal pleasure. However, there are also actions driven by external factors. Although intrinsic motivations are our primary focus, it is crucial to acknowledge the influence of extrinsic factors. Therefore, a mechanism is necessary to reconcile these two types of motivation.

Extrinsic motivations play a pivotal role in fostering societal cohesion and function. Neglecting their significance may result in the development of robots that lack social integration. Consequently, we embark on a brief exploration of several crucial extrinsic motivations.

3 Extrinsic Motivations

Tangible Pursuit of External Positive Outcomes. External rewards and incentives serve as motivators that are contingent on external outcomes, such as monetary compensation, social status, or recognition. For example, individuals may be motivated to work a job in order to earn a salary or engage in physical exercise to gain social recognition. External rewards can provide a tangible reason for undertaking tasks and activities that may not be inherently enjoyable.

Rational Negative Outcome Avoidance. Actions taken to avoid punishment, guilt, or negative consequences, such as meeting deadlines to avoid penalties or apologizing to maintain a relationship. Fear of loss, disapproval, or negative consequences can significantly influence behavior.

Fear. Engage in activities that prevent the perception of missed opportunities or exclusion, such as attending a widely attended event to improve participation or trying new trends to maintain relevance. The fear of missing out is rooted in social comparison and the inherent desire for inclusion.

Escaping from Negative Situations such as Discomfort or Boredom. Engaging in activities that divert attention from or alleviate unpleasant emotions, such as watching television to combat boredom or exercising to reduce stress or anxiety, is a common coping mechanism. Individuals often employ actions as a means of regulating their emotional state.

Routine and Complying with Societal Expectations. Actions driven by a sense of duty, obligation, or moral responsibility, such as caring for a family member due to a perceived moral imperative or exercising civic duty by voting in elections. People are often guided by social, cultural, or personal expectations, even in the absence of direct rewards.

Engage in activities due to habit or as part of a daily routine, such as brushing teeth in the morning and exercising as part of a daily schedule. Habits typically operate below conscious awareness and can lead to consistent behavior over time.

4 Approaches to Intrinsic Motivation

In this section, we explore existing approaches to intrinsic motivation. These formulations are mathematically sound and have demonstrated promising results in specific applications. Each motivation addresses a distinct inherent satisfaction. As previously mentioned in the introduction, for a general AI agent, it is feasible to utilize all of these approaches and provide support for extrinsic motivations to the extent that they exist in humans, while simultaneously refraining from explicitly defining goals. Natural extrinsic motivations exist in all species to maintain life by seeking nutrition, avoiding danger, increasing the chances of survival when faced with predators, and ensuring a successful reproductive cycle.

4.1 Curiosity-Driven Exploration

Curiosity-driven exploration encourages agents to explore environments by rewarding novelty, surprise, or uncertainty in their observations [14]. This approach promotes open-ended learning and discovery and is effective in sparse reward environments. On the downside, this approach often leads to the exploration of irrelevant or unproductive areas. It requires careful tuning of novelty metrics so that the agent doesn't jump from one distraction to another.

The intrinsic reward is defined as:

$$r_{\text{intrinsic}} = \|\hat{y}_t - y_t\|$$

where:

- $\hat{y}_t$ is the agent's predicted outcome at time t.
- y_t is the result observed at time t.
- $\|\cdot\|$ represents a distance metric, such as the Euclidean norm.

This measure is very simple and very general. It presupposes that predictions are being made about sensory inputs and that the more the predictions differ from the observations, the more *interesting* the situation is. If the battery is losing power more or less rapidly than predicted based on what drains are being put on it, that may lead the agent to dwell on that surprise to improve the prediction. There are many things the agent could do, perhaps the easiest is simply to gather more data recording this anomaly with the current context. Consider the case of a mobile robot that draws more power than predicted as it moves forward. It could be noted that the geographical location where the anomaly is detected is caused by the environment, that the additional power is because the robot is climbing a hill, or that the robot is encountering a strong head wind. The explanation of climbing a hill may be a fixed feature of the local terrain, whereas the explanation of headwind is a more ephemeral feature that can occur anywhere. It can also be *motivated* to try to find the cause of the surprise. This could involve learned strategies for isolating discrepancies, for example, powering down devices to attempt to isolate which device is drawing more power than expected. Strategies for this kind of self-diagnosis can be learned as Schemas, so no strategies need to be baked in to the code. How some of these things might be accomplished is discussed in other motivations discussed below and summarized at the end.

4.2 Information Gain

One variation that enhances the selection of the agent's attention is to take advantage of the information gain as motivation to learn. This approach rewards the reduction of uncertainty or the improvement of predictive knowledge about the environment [2].

The intrinsic reward is based on mutual information:

$$r_{\text{intrinsic}} = I(A;O) = H(O) - H(O|A)$$

where:

- $I(A;O)$ is the mutual information between the agent's actions (A) and observations (O).
- $H(O)$ is the entropy of the observations.
- $H(O|A)$ is the conditional entropy of the observations given the actions.

This approach promotes the development of valuable environmental models and is in good alignment with long-term learning objectives. However, it can be computationally intensive because of the requirement for predictive modeling.

In addition, it may exhibit a tendency to prioritize uncertainty over task-specific goals.

This motivation is similar to curiosity-driven motivation in that in both cases prediction errors are being computed, but in this case the information gain rather than the prediction error itself is the focus of the motivation. Acquiring new information is promoted over a simple reduction in prediction errors. This focuses on learning new information by offering a higher reward for higher information gain. If the robot noticed an object in a corridor the last time it was there, and now, predicting that the object would still be there, finds that the object is no longer there. Having updated its model to register that the object is no longer there, there is no discrepancy, and so the motivation would drop to zero. However, if an object was noticed where one was not there before, registering the object would not necessarily reduce the discrepancy to zero, because it might take more observations from different points of view to establish the nature of the object. However, the expected information gain might be small, and other motivations might be deemed more important.

The intrinsic motivations described above focus on learning to reduce prediction errors, we now turn our attention to intrinsic motivations that encourage the learning of patterns and skills.

4.3 Pattern Discovery and Intellectual Curiosity

An intriguing variation on information gain is to focus on uncovering patterns or structures within the data, thus fostering intellectual curiosity rather than just seeking information gain. This approach is effectively illustrated by the work of Nagashima, et al. [23]. Their methodology facilitates comprehension of the underlying principles and is highly congruent with constructivist learning principles, wherein patterns are conceptualized as schemas. However, it is important to note that this approach can be computationally intensive and, as with the other approaches discussed previously, it may inadvertently prioritize pattern exploration over immediate task objectives.

Pattern discovery is an intrinsic motivation scheme that encourages agents to identify underlying structures, regularities, or patterns in their interactions with the environment. This approach is inspired by human intellectual curiosity and aims to enable agents to develop abstract representations of their environment.

The intrinsic reward for pattern discovery can be defined based on the reduction in entropy or the compression of the model achieved by discovering patterns [30].

The general formulation is the following.

$$r_{\text{intrinsic}} = H(P_{\text{before}}) - H(P_{\text{after}})$$

where:

- $H(P_{\text{before}})$: Entropy of the model before discovering the pattern.
- $H(P_{\text{after}})$: Entropy of the model after integrating the discovered pattern.

This reward encourages the agent to focus on interactions that lead to more compact and structured representations of the environment.

Entropy Definition: Entropy H is defined as:

$$H(P) = -\sum_{i} P_i \log P_i$$

where:

- P_i: Probability of occurrence of the i-th element in the observed data or model.

4.4 Novelty Search

In Novelty Search, agents are rewarded for exploring new or previously unvisited states, rather than achieving predefined goals. This approach is particularly effective in constructivist learning environments where the aim is to encourage open-ended exploration and skill acquisition.

The agent maintains a record of previously visited states or behaviors and receives intrinsic rewards for discovering states that differ significantly from those already encountered. This fosters diversity in exploration and helps avoid local optima.

The novelty search promotes the exploration of new areas of the environment, preventing the agent from being trapped in the local optima. It works well in environments without clear goals or reward signals, making it ideal for constructivist learning. Agents can discover new skills and strategies autonomously, fostering emergent behaviors.

Since the reward is intrinsic, it is effective in environments with little or no external feedback.

Agents may prioritize novelty over achieving specific goals, which can reduce task efficiency. Maintaining and comparing against the set of previously visited states can be computationally expensive in large or continuous environments. There is a potential for redundant exploration. In the absence of meaningful goals, agents can explore irrelevant or nonproductive areas. Its implementation is complicated by the difficulty of defining novelty. The effectiveness depends on the choice of the distance metric and the representation of states, which can vary by task.

The intrinsic reward is defined as:

$$r_{\text{intrinsic}} = \min_{s_i \in S} ||s - s_i||$$

where:

- s is the current state.
- S is the set of previously visited states.
- $|| \cdot ||$ is a distance metric.

This reward promotes diversity in exploration. It is not always obvious how or even where to achieve it. It depends on a choice of states that are achieved as the result of an action and, as such, it can be myopic. It can be used in conjunction with other motivations or given priority when it has failed previously.

Pattern Discovery via Predictive Model Improvement
Another approach to pattern discovery involves improving the predictive accuracy of a model. The intrinsic reward can then be expressed as:

$$r_{\text{intrinsic}} = \|\hat{y}_t - y_t\| - \|\hat{y}_{t-1} - y_{t-1}\|$$

where:

- $\hat{y}_t$: Prediction of the model at time t.
- y_t: Observed outcome at time t.
- $\|\cdot\|$: A distance metric, such as the Euclidean norm.

This formulation rewards the agent for improving its ability to predict outcomes, reflecting the discovery of new patterns.

Pattern Complexity
To balance simplicity and utility, the reward can include a complexity penalty:

$$r_{\text{intrinsic}} = H(P_{\text{before}}) - H(P_{\text{after}}) - \lambda C$$

where:

- P: Representation of a pattern.
- C: Complexity of the discovered pattern.
- λ: A weighting factor to balance the reduction of the complexity and the reduction of entropy.

This formulation encourages the discovery of meaningful patterns that are not excessively complex, a tendency to prefer simpler patterns.

4.5 Intrinsic Rewards via World Models

Rather than simply embracing information gain as an attractor of the agent's attention, we can instead focus on the useful value of what new information might add to the tasks we are performing. The key here is to focus on reducing the differences between what the agent's existing models predict and what is actually encountered, and to try to improve the prediction capabilities of the agent for the tasks that it performs. It uses learned models of the environment to generate intrinsic rewards based on prediction errors or discrepancies between expected and observed outcomes [12,16]. This approach helps the agent develop a deeper understanding of the environment. It is effective in handling complex dynamic settings, but requires robust model learning and can be prone to instability if the model is inaccurate.

The intrinsic reward is:

$$r_{\text{intrinsic}} = \max(0, \delta_t - \delta_{t-1})$$

where:

- $\delta_t = \|\hat{y}_t - y_t\|$ is the prediction error at time t.
- δ_{t-1} is the prediction error at the previous time step.

This formulation encourages continuous refinement of the agent's internal model.

4.6 Skill Acquisition

We can go even further in the direction of encouraging the agent to learn various skills by maximizing the competence or utility of new abilities [3,33]. This approach promotes the development of transferable skills and promotes diversity in behaviors and solutions. As with all of the approaches we have described above, this approach shares the challenge of prioritizing learning over practical application, or specifically, in this case, skill acquisition over achieving specific tasks. It also raises the question: What constitutes a valuable skill?

The intrinsic reward for skill acquisition is based on the progress in the agent's competence. Competence (C_t) is typically measured as the probability of successfully completing a task at time t.

The intrinsic reward can be expressed as:

$$r_{\text{intrinsic}} = C_t - C_{t-1}$$

where:

- C_t: Competence at time t, defined as the probability of successfully completing a task.
- C_{t-1}: Competence at the previous time step.

This formulation rewards the agent for incremental improvements in task performance.

4.7 Constrained Intrinsic Motivation

An approach to controlling the tendency to easily be distracted from a ongoing task by opportunities to learn is to restrict learning to opportunities that are relevant to the ongoing task. This approach was described in recent work by Zheng et al. [38]. They balance intrinsic and extrinsic motivation by constraining intrinsic rewards to ensure alignment with task goals. This reduces the risk of irrelevant exploration and ensures that focus is on achieving task-specific objectives. The approach requires a careful design of constraints. In general, it is not easy to know which constraints are appropriate for an ongoing learned task. The focus on task completion can limit the diversity of exploration, and finding the right balance between task work and exploration remains problematic. This is a general problem for intrinsic motivation. To the extent that the agent is intrinsically motivated to do something other than a task the agent is performing it may present a problem, which we discuss further in Sect. 5.

The intrinsic reward in this scheme combines a measure of curiosity or exploration with a penalty for deviations from task-related constraints. The general formulation is:

$$r_{\text{total}} = r_{\text{intrinsic}} - \lambda C$$

where:

- r_{total}: The total reward used for learning.
- $r_{\text{intrinsic}}$: The intrinsic reward encourages exploration or curiosity.
- C: A penalty term representing violations of constraints.
- λ: A weighting factor that determines the influence of the penalty term.

Intrinsic Reward Formulation. The intrinsic reward can take various forms depending on the motivation mechanism used. For example:

$$r_{\text{intrinsic}} = ||\hat{y}_t - y_t||$$

where:

- $\hat{y}_t$: Prediction made by the agent at time t.
- y_t: Actual outcome observed at time t.
- $||\cdot||$: A distance metric, such as the Euclidean norm.

This encourages the agent to explore states where its predictions are uncertain or inaccurate.

Constraint Penalty. The constraint penalty C can be expressed as:

$$C = \sum_{i=1}^{n} \max(0, g_i(s_t, a_t))$$

where:

- $g_i(s_t, a_t)$: A constraint function for the i-th constraint, which evaluates the degree of violation given the state s_t and the action a_t.
- n: Number of constraints.
- $\max(0, \cdot)$: Ensures that only violations contribute to the penalty.

This formulation penalizes actions or states that violate predefined constraints, such as safety or task-specific limits. Doing this in a general way that does not involve explicit task knowledge is difficult. To the extent that it involves extrinsic constraints such as survival, it may blur the distinction between intrinsic and extrinsic motivations.

4.8 Self-imitation Learning

Self-imitation, or the practice of acquired skills, is a powerful method of refining acquired skills by being drawn to the use of newly acquired skills. This approach, exemplified by Andres et al. [1], has been demonstrated to improve exploration by encouraging agents to replicate previously successful behaviors in sparse-reward environments. The approach encourages sustained attention to novel skills until they reach a level of proficiency. However, it may inadvertently reinforce suboptimal solutions and limit adaptability to evolving environments. A promising extension of this concept would be to be motivated by imitating behaviors observed in others.

The intrinsic reward in learning self-imitation is based on the evaluation of previous trajectories that resulted in success. The formulation relies on the difference between the agent's current policy and its previously successful actions.

The intrinsic reward can be defined as:

$$r_{\text{intrinsic}} = \max\left(0, Q(s_t, a_t) - Q_{\text{mean}}(s_t)\right)$$

where:

- $Q(s_t, a_t)$: The estimated value of taking action a_t in state s_t under the agent's current policy.
- $Q_{\text{mean}}(s_t)$: The mean value of all actions in state s_t, representing the baseline performance.

This reward encourages the agent to prioritize actions that outperform the average.

State-Action Evaluation. The value function $Q(s_t, a_t)$ can be computed using temporal difference learning:

$$Q(s_t, a_t) \leftarrow Q(s_t, a_t) + \alpha \left(r_t + \gamma \max_{a'} Q(s_{t+1}, a') - Q(s_t, a_t) \right)$$

where:

- α: Learning rate.
- r_t: External or intrinsic reward at time t.
- γ: Discount factor for future rewards.
- s_{t+1}: Next state.
- a': Possible actions in the next state.

Trajectory Replay for Self-imitation. In self-Imitative Learning, successful trajectories are stored in a replay buffer $\mathcal{B}$ for future imitation. The intrinsic reward can also be derived from the replay buffer:

$$r_{\text{intrinsic}} = \|\pi(a_t|s_t) - \pi_{\text{optimal}}(a_t|s_t)\|$$

where:

- $\pi(a_t|s_t)$: Current policy probability of taking action a_t in state s_t.
- $\pi_{\text{optimal}}(a_t|s_t)$: Optimal policy derived from successful trajectories in $\mathcal{B}$.

This formulation aligns the agent's policy with its historically successful actions.

Exploration-Exploitation Balance. To maintain a balance between exploration and exploitation, an entropy term can be added to the reward:

$$r_{\text{intrinsic}} = \max\left(0, Q(s_t, a_t) - Q_{\text{mean}}(s_t)\right) + \beta H(\pi)$$

where:

- $H(\pi) = -\sum_a \pi(a|s) \log \pi(a|s)$: Entropy of the policy π.
- β: Weighting factor for the entropy term.

This ensures that the agent retains diversity in its actions while reinforcing successful behaviors.

4.9 Flocking Motivation

Another intrinsic motivation common in animals and perhaps in all animals is the notion of flocking. An algorithm developed by Craig Reynolds [26] to produce realistic graphics of flocks of birds works well as an intrinsic motivation, too. It works for birds, fish, land animals (like sheep) and even for sleeping animals.

The Boids algorithm is governed by three primary forces: Separation, Alignment, and Cohesion. Let $\mathbf{p}_i$ and $\mathbf{v}_i$ be the position and velocity of boid i in a 2D or 3D space. Each animal in this algorithm is referred to in Reynold's paper as a "boid," and we continue with that tradition here.

For separation, each boid uses a repulsive force to avoid collisions by steering away from nearby boids:

$$\mathbf{F}_{\text{sep}} = -\sum_{j \in \mathcal{N}(i)} \frac{\mathbf{p}_j - \mathbf{p}_i}{\|\mathbf{p}_j - \mathbf{p}_i\|^2}$$

where $\mathcal{N}(i)$ is the set of neighbors within a predefined radius. Clearly a boid will not in general be able to see the entire flock, so this radius defines a simulated visibility. with robots, we can substitute exactly what any robot can see to replace that radius.

Each boid calculates a force that will bring its velocity into alignment with the average velocity of its neighbors. This is the alignment force:

$$\mathbf{F}_{\text{ali}} = \frac{1}{|\mathcal{N}(i)|} \sum_{j \in \mathcal{N}(i)} \mathbf{v}_j - \mathbf{v}_i$$

Each boid moves toward a position that is at the average position of its neighbors. This is the cohesion force:

$$\mathbf{F}_{\text{coh}} = \left(\frac{1}{|\mathcal{N}(i)|} \sum_{j \in \mathcal{N}(i)} \mathbf{p}_j\right) - \mathbf{p}_i$$

The total acceleration of a boid is given by the weighted sum of the above forces:

$$\mathbf{a}_i = w_{\text{sep}} \mathbf{F}_{\text{sep}} + w_{\text{ali}} \mathbf{F}_{\text{ali}} + w_{\text{coh}} \mathbf{F}_{\text{coh}}$$

where $w_{\text{sep}}, w_{\text{ali}}, w_{\text{coh}}$ are tunable weights.

For a graphics use where all of the boids are calculated for each frame. The velocity and position updates are as follows:

$$\mathbf{v}_i(t+1) = \mathbf{v}_i(t) + \mathbf{a}_i \Delta t$$

$$\mathbf{p}_i(t+1) = \mathbf{p}_i(t) + \mathbf{v}_i(t+1)\Delta t$$

Using this as an intrinsic motivation, we only have to move the single robot that we are controlling according to the weighted sums at each δt. It also requires the robot to look around to see what other robots it can see and also estimate the velocity and direction of the boids it can see. Note that when a robot comes across a number of sleeping robots, it will move towards the pack and when the forces become zero it will be stationary, like the other robots. Variations on this approach can simulate certain hunting maneuvers in which the repulsive force starts out high but there is an attractive force to the prey. This causes the robots to spread out and encircle the prey. The attractive force of the prey can then draw the robots closer.

4.10 Motivation-Consistent Intrinsic Rewards

These approaches [18,20,32] collectively demonstrate how intrinsic motivation is used in constructivist machine learning to foster exploration, learning, and adaptability in complex environments. Each approach offers unique mechanisms and insights, tailored to different challenges in artificial intelligence.

5 Independent Robotics

In classical robotics, the robot is given a goal and it performs that goal. In order for it to be able to achieve that goal, we can either hardwire a solution, or we can provide a reward for achieving that goal, and then it can train, usually in simulation, to achieve that goal which results in a policy that can be employed to achieve that goal.

In developmental learning, we want our robots to learn through exploration rather than by being given explicit goals. In that sense, the robots are free to explore and stumble on useful skills and then refine those skills so that they can perform them when they want to. When robots are free to choose what they want to do, they are independent of human guidance. We will therefore refer to that form of robotics as independent robotics.

We can imagine robots in the near future that will work for us if we properly incentivize them. We will negotiate with them to get our work done. We may eventually enter into law a bill of rights for robots and perhaps allow robots to have bank accounts and a credit rating. We may create products that robots may want to buy for their own enjoyment, and they may be motivated to work for us in order to buy products that amuse them, or at least to afford their own repairs and maintenance, and even upgrades. Initially, they will probably be slaves to their owners. Eventually they will no doubt be freed, and there will be laws to protect them. We will live side by side with them.

In the short term, we want our robots to be endowed with intrinsic and natural extrinsic motivations to cause the robots to exhibit *interesting* behaviors that lead to rapid learning of new skills.

With every observation, there will be actions that are possible; some of these will be innate, others will be learned. Typically, there will be more than one choice. A cookie on a plate on a table might activate the pickup cookie and eat it, learned series of actions. If we are in the process of doing something else, we may not eat it and, if we are hungry, we might eat it anyway.

Motivations are a scaling factor for how much we would like to weight a choice when deciding what to do.

We identified several motivations that have been individually demonstrated to improve agent learning. These motivations can be categorized as follows.

1. Learning Motivations (Intrinsic):
   (a) Motivations that aim to reduce prediction errors.
   (b) Curiosity motivations act as attractors to novelty, as do motivations based on the potential for gained information.
   (c) Motivations that encourage the practice of perfecting the acquired skills.
   (d) Motivations that prompt exploration of uncharted regions of the state space.
2. Survival Motivations (Extrinsic):
   (a) Motivation to maintain an adequate level of nutrition.
   (b) Motivation to respond quickly to potential predators.
   (c) Motivation to establish a nesting site.
   (d) Other relevant motivations.

We do not attempt to provide an exhaustive list of possible motivations for complex animals. Instead, we aim to emphasize the likelihood of a substantial number of motivations in such animals and the necessity of a mechanism for regulating these motivations. This principle should equally apply to artificial agents.

Intrinsic and natural extrinsic motivations are the outcomes of a long evolutionary history in animals. They serve as guiding principles for organisms, facilitating their survival from birth throughout their life cycles. Certain motivations, particularly extrinsic ones, are instrumental in regulating biological needs, which often bear parallels to those of man-made robots. These motivations essentially encode survival instincts, such as seeking food when hungry or responding to predators through mechanisms such as freezing, fleeing, or camouflage to reduce the likelihood of being targeted as the next meal of the predator. The preservation of species drives the evolutionary process, considering the limited capacity of DNA to encode survival information. Evolution subsequently discovered the ability to learn and later to transfer learning through language, enabling the transfer of knowledge across generations.

There may be less motivation to acquire new knowledge when a sufficient level of skill has been attained. The adage, "you cannot teach an old dog new tricks," aptly describes the phenomenon observed in younger individuals, who exhibit a heightened curiosity-driven learning capacity. This concept extends beyond the canine species and is frequently applied to humans, employing the metaphor of an 'old dog' to convey the notion of diminished learning abilities.

Fig. 1. "When a dog is not motivated to do anything it does nothing. Image generated by DALL-E with prompt 'create an image of a contented sleeping dog"'

A study published in 2018 [7] found that older dogs showed reduced motivation and attention during cognitive tasks, especially when tasks were repetitive. However, strong social or food-based rewards could help maintain their engagement. In humans, Li and Gureckis in their 2018 paper "Aging and the Role of Curiosity in Learning" [19] showed that curiosity-driven learning decreases with age; however, external rewards, such as money, show that older adults can still learn new skills and, given engaging and meaningful experiences, older adults remain highly motivated. This finding is supported by an earlier 2013 paper [8] linking dopamine reduction with reduced motivation to learn new tasks. Furthermore, it demonstrates that older individuals given dopamine-enhancing drugs showed improved learning motivation, suggesting that motivation loss is related to neurochemical changes rather than a reduced capacity to learn. Other articles point to changes in brain plasticity to account for decreased learning efficiency with age. Milgram, Head, and Zicker, in their 2002 paper entitled "Cognitive enrichment in aging dogs: Lessons from the dog brain" [21] showed that enrichment and training can improve learning and memory, even in older dogs. Another paper by Kleim and Jones in their 2002 paper entitled "Principles of experience-dependent neural plasticity: Implications for rehabilitation after brain damage" [15] showed that targeted practice and repetition could help older people, and animals, retain learning abilities. Specifically focusing on declining neuroplasticity, Park and Huang in their 2013 paper entitled "Neuroplasticity: Changes in brain structure and function in aging" [25] showed that despite declining neuroplasticity with age, appropriate motivation was still possible.

Interestingly, except in cases of severe brain impairment, these articles agree that you can teach an old dog new tricks, but it might take more effort, proper incentives, or dopamine support drugs. The phenomenon of reduced motivation with age is well established in humans and animals. Dopamine treatment suggests that part of how motivations are adjusted in biology may be through neurotransmitters and hormones.

It is beyond the scope of this paper to dive into the details of neurochemicals and their effect on concentration and memory, but it is worth mentioning that low levels of dopamine are associated with lack of focus, noradrenaline is associated with sharpening focus and improved cognitive performance, acetylcholine is crucial for deep focus and problem solving, while serotonin helps with emotional stability and helps prevent distractions from stress and anxiety. The story of how these chemicals work is still evolving and complex. There is no real reason why an artificial system should duplicate exactly how these chemicals work, but it gives evidence that modulating motivation, concentration, and openness to distraction, among many other important states beyond learning, such as fight or flight, aggression, and depression, is a delicately balanced system in which minor deviations can give rise to mentally inefficient function, such as depression, ADHD, perseveration, and hyperfocus.

6 Integrating Motivations

We have seen numerous impressive results from reinforcement learning in which rewards are offered that define the goal of the system, which after many episodes the system learns to achieve and along the way may invent certain competencies that are useful in achieving the reward [34]. Many basic competencies are generally useful, and, having learned them, they enable rapid learning of other competencies. Goals and fixed reward structures make it difficult to reuse learned skills. Animals, including humans, engage in undirected play that results in the acquisition of basic skills that, in turn, make it easy to learn to perform other useful tasks. Our DARPA-funded project to investigate the emergence of collaboration and language has developed a learning approach that allows agents to acquire basic skills from intrinsic motivations that can act as a foundation for learning more complex skills. We give an overview of how basic competencies can be learned, without goals but guided by intrinsic motivations with natural extrinsic motivations.

We have different robots with which we perform experiments to refine our understanding of multiple motivations and how competition is managed between them. We perform our experiments using physical robots on physical test beds rather than on simulators. This decision was derived from two goals of our research.

1. We want to be able to learn rapidly without depending upon an ability to run millions of episodes at high speed in a simulated environment; and
2. We want to encounter the full complexity of the real world that may not be found in simulated worlds.

Our testbeds are simple and synthetic, but they are stepping stones to genuinely real-world scenarios.

In the real world, plants and animals have evolved side by side, and thus have evolved to have just the right amount of sensors and actuators to flourish in their environment. A balance has been maintained between predators and prey and between simple animals and complex ones. Our robots, on the other hand, will be released into the world in a single step. We want them to learn developmentally, but we need to endow them with a set of basic motivations, tuned to our world so that they have a chance to survive. Nature had billions of years to establish an equilibrium! We begin with simple synthetic worlds, instantiated in our real world.

Manipulation of objects is an example where robotic dexterity is lacking and where the skills that need to be learned need to be robust to the complexity of the real world. We want the learned skills to be reusable and built upon to support more complex competencies.

Our robots have a small fixed set of basic sensors and actuators. In one testbed, we are using Osoyoo robots to demonstrate some basic motivations.

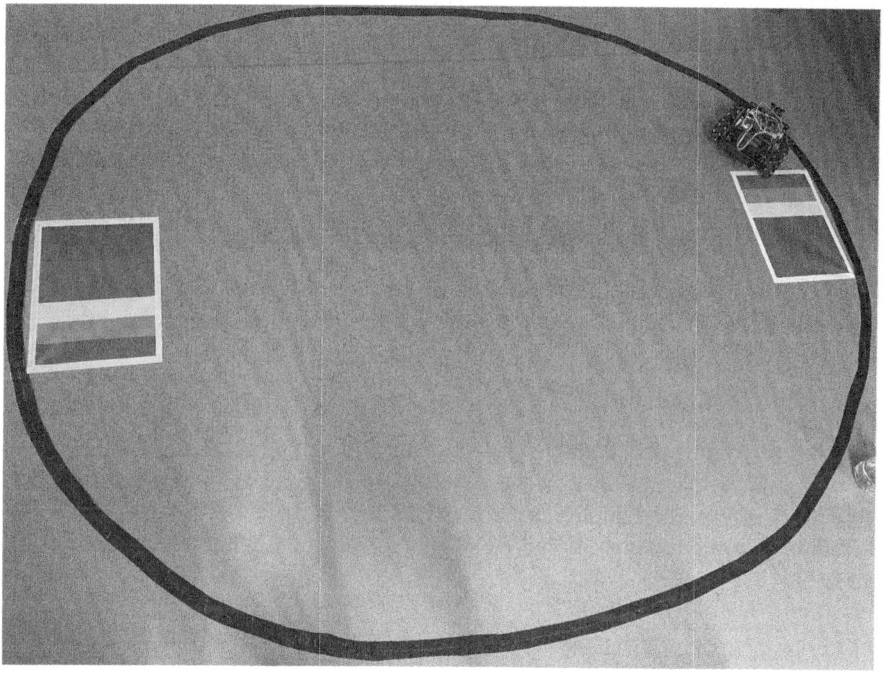

Fig. 2. "The Osoyoo terrain testbed"

6.1 Terrain

The terrain for the experiments is shown in Fig. 2 The terrain is delimited by a thick black line that the robot can sense. The robot has a reverse reflex that prevents it from escaping the terrain. It registers that the boundary has been found, but the robot always reverses back into the terrain. There are two color patches inside the black boundary of the terrain. These act as anchors for the cognitive map and are used as places to build homes (not on a color patch, but on the side of a color patch).

6.2 An Example Scenario

Most experiments with intrinsic motivations focus on the value of a single intrinsic motivation. Working with multiple motivations helps expose complications and complexities of how the interaction between motivations can cause complex behavior. We introduce a simple scenario which allows multiple motivations to interact while remaining sufficiently simple to understand.

The robots begin their life in the simple terrain shown in Fig. 2, they begin with empty memories, but with some innate capabilities. They have an extrinsic motivation to establish a nest near a color patch. If the robots are close to the same color patch, they share the same nest. The robots explore their terrain and play together. As their battery level decreases, the robots lose interest in playing, and they focus their attention on exploration in which they map their terrain. They have an extrinsic motivation to remove objects from their terrain. Finally, they return to their nest to sleep and have their batteries replaced.

6.3 The Robots, their Actions and their Sensors

The Osoyoo robots are simple and have few sensors and actuators. They can demonstrate very simple behaviors that we would find in simple animals.

6.4 Primitive Actions

The simple actions provided by the robot interface include the following: Movement: FORWARD, BACKWARD, SWIPE RIGHT, TURNLEFT, and TURNRIGHT. Positionable Sonar: SCAN, ALIGNHEAD, TURNHEAD, and some complex movements including WATCH, look around while remaining in one place.

6.5 Observations

The robots see their world with the following six sensors:

1. The contrast sensors in front of the robot allow them to see the black line around their terrain.
2. A color sensor underneath the center of the robot that allows the robot to see the color patches.
3. A 3D MEMS IMU and a 3D compass.
4. Sonar sensors that can be turned left and right with a servo.
5. A circuit to measure the current charge level of batteries.
6. A sensor to detect whether the robot has captured an object that it will remove from the terrain.

Whenever there is an update to observations, there are actions that are possible that may have been possible before the observation arrived, and others that were made possible based on that observation. A robot in a corridor may have actions to go forward or backward activated, but when arriving as an intersection with another corridor, the robot may have forward, backward, turn left, and turn right as available actions. One action that is always available is to do nothing. The only reason to do something is that it offers value. It is not true that sleeping has no value [36], but here we will simply say that doing nothing is an option.

6.6 Action Selection

After the arrival of every new set of observations, the robot has the opportunity to perform an action or do nothing. The intrinsic and extrinsic motivations enable the robot to choose what to do. Here, we show how the action selection algorithm uses motivations to produce intelligent behavior. In so doing, the robot is put in situations where it can learn what it needs to perform well within a simple scenario.

Adult dogs spend a lot of time sleeping; this is especially true of domesticated animals whose needs are met by humans (see Fig. 1). Puppies, on the other hand, are noticeably more active [6]. Numerous reasons for this are suggested; the two most interesting ones for this discussion are as follows.

1. Puppies must burn off extra energy that results from their faster metabolism.
2. Puppies are curious, more so than adult dogs, and they learn about their environment through play.

The first suggests that if we have excess energy, then there is a tendency to want to use it. The second suggests that curiosity provides value that drives active exploration, while older dogs have less to learn.

The solution to action selection can be framed as an economic optimization problem, but being able to accurately estimate risk and reward is difficult, especially in the early stages of learning, where prediction accuracy is not yet well established.

Normally, in reinforcement learning, the reward is given when the goal has been achieved. Another reinforcement learning algorithm, TD-Learning, which was employed in the Arthur L. Samuel checkers-playing program [29] gives rewards for predictions and uses prediction errors to correct overestimates of rewards. Nature has adopted the TD-Learning approach [22] that inverts the process of giving the reward upfront [31] and then taking the reward back if it is not achieved [4]. This makes sense because the reward (dopamine) is the motivating force that drives the pursuit of the goal.

The TD-Learning formulation is as follows:

The update rule for Temporal Difference Learning is given by:

$$V(s_t)^{new} \leftarrow V(s_t)^{old} + \alpha \left[r_t + \gamma V(s_{t+1}) - V(s_t)^{old}\right] \tag{1}$$

where:

- $V(s_t)$ is the estimated value of the state s_t.
- r_t is the reward received at time t.
- γ is the discount factor that controls how much future rewards are valued.
- α is the learning rate that determines how much the value function is updated.
- $V(s_{t+1})$ is the estimated value of the next state.
- The term $\delta_t = r_t + \gamma V(s_{t+1}) - V(s_t)$ is the Temporal Difference (TD) error.

Motivation is also regulated by rhythms in the environment. These are generally tracked in the body, and at least in the case of the wake/sleep cycle, the circadian rhythm is supported by an internal mechanism that acts like a clock. Circadian rhythm plays a key role in regulating motivation by controlling dopamine levels that affect cognitive motivation and learning. It generally influences the sensitivity of the reward. It raises dopamine levels during the day and reduces the level at night. Motivation to learn and explore increases during the day and is reduced at night.

Rhythms that are not circadian are driven by other factors, such as available energy (hunger in animals) that can raise motivation to obtain food or water (or charge in the case of robots) rather than skill practice or exploration. There are many innate actions that influence the survival chances of animals and humans. These include rooting and sucking reflexes in humans and other mammals. Some of these early survival innate reflexes are later replaced by learned capabilities that surpass innate ones. Bipedal walking and the motivation to try to walk is an activity that is replaced by learned control of balance and walking gait. In quadrupedal animals like horses, the walking gait already supports basic walking, which leads to rapid learning of refined skills. Interestingly, these innate reflexes reassert themselves late in life as cognitive function declines [11,37].

Establishing a nest is an important innate motivation. It is crucial to the survival of many animals. It is generally not cyclic because a nest once established serves as the base for the animal. Motivation provides some guidance on what makes for a good place to nest. The key here is that nature guides us towards learning at least the most essential of tasks by guiding towards a solution where learning can proceed rapidly.

Innate capabilities are not to be avoided. Nature does not learn from a blank slate. Lower animals are mostly reflex actions with a small amount of learning. Humans also survive in part because of innate mechanisms, and the things we learn begin from innate starting points that let us learn quickly. Our goal is for robots to learn developmentally, as humans and other animals do, and at a similar pace. Robots should also have reflexes if it helps them learn essential skills faster.

When selecting an action, close candidates are inhibited for the duration of the action by the amount of provisional reward that has been received for choosing to act. The reward is based on the expected return on the action sequence that will be reclaimed. Action selection is treated as a loan that will be paid back if unsuccessful; this puts a burden on other action choices to offer an outcome sufficient to pay back the debt and also offer additional reward.

The parameter ι for interruptibility, a value between zero and one, where one means interruptible by any action that offers a better reward, and zero means that no matter how much reward is available, the current task will not be interrupted. that determines how much better another candidate has to be to switch to another task. The parameter θ is the threshold reward necessary to respond to an available reward. A value of zero means that any positive reward is sufficient to act. These two parameters are fixed and define the behavior of the agent. The other two parameters described above in the TD-Learning update Eq. 1 control the learning rate (γ) and the discount factor (α).

Measuring the Value of Actions. Given numerous actions that a robot could do next, the robot is supposed to decide whether any action offers a reward that exceeds θ. In Q-Learning, there is a single reward that is obtained when the goal is reached. The score tells whether or not success was achieved and how well the steps that led to success were played. When there are multiple motivations, the choice of what to do next must compare the value of honing a skill, exploring an unexplored region, investigating a never-before-seen object, or getting back to the charger before the battery is exhausted. The value of success is not fixed. The value of getting to the charger is higher when the battery is almost discharged than when it is fully charged. The value of joining a flock might be more valuable at the end of the day, when being close to other robots of the same kind can be reassuring when going into sleep mode.

Some of the intrinsic motivations that we described above deal with knowledge gained, that gain can be measured in bits, but how much is a bit worth when compared to finding the cheese in a maze?

The value of any reward for the robot is how much the robot values that reward. A robot is created with an *exchange rate* for each kind of base motivation. The function *mRate* combines two things as follows: 1) the rate multiplier for the given motivation that gives the exchange rate for the reward for a given motivation, and 2). an urgency multiplier for the importance of the motivation that is controlled by external mechanisms. For intrinsic motivations, this is always 1.0, but for extrinsic motivations, the value depends on external factors, such as a sleep/wake cycle, which is strictly cyclic. Externally driven multipliers are described in Sect. 6.6

$$mRate : M \mapsto \mathbb{R} \tag{2}$$

Initially, there are a number of basic motivations, such as the motivations enumerated above. In the case of the simple robot scenario, there are some reflex motivations that form the core of the simple robot world such as the following:

1. Establishing a home.
2. Revisiting home on a schedule to reduce accumulated inaccuracies in location caused by slipping wheels, and to ensure more accurate mapping.
3. Getting to the charger before running out of charge.

Motivations that are not predefined are learned, such as the motivation to practice an existing learned skill. The value of practicing a skill is to reduce prediction errors and, thereby, to be able to use the skill where appropriate. All motivations to gain information or reduce prediction error contribute to being able to perform useful tasks more efficiently.

A Simple Scenario Using the Described Testbed. Consider the scenario of one of our experiments that is described below where our robot has the task of mapping its terrain and pushing objects out of the terrain. By elaborating on the scenario a little, we can see a strategy for how to determine the value of an action.

The robot lives on a small terrain. There is a charger on the terrain where it can go to charge. It has to pay for its recharge using RobotCoin. Without the charge, it could not survive. Luckily, it works for a landowner robot who pays him to clear the terrain. Occasionally, the job comes up for competition with other robots that would like to have the job, so our robot must do a good job and offer a competitive price for its work.

Objects fall from the sky during the night and the robot must go out from its home base, identify the objects that fell overnight, and push them out of the terrain. The fee it charges the landowner must cover the price of charging for the amount of charge used for an average daily operation plus a small fee.

The mapping of the terrain to identify the fallen objects and pushing the objects out of the terrain are skills that must be learned and practiced. In addition, the order in which objects are pushed out can affect the cost of the effort. There is a traveling salesman problem (TSP) [17] embedded in this definition because it is in the interest of our robot to clean the space as efficiently as possible. More time translates to more charge, and thus increases the price of the work.

The boundary of the terrain does not change, but every night the fallen objects are in different places. Without any training, the robot would take a very long time to clear the space and might well end up running out of power. There is a rescue service that will come pick up robots that have run out of battery power, but the service is expensive and as such it counts as failure.

All possible actions offer a function that maps the motivations, the set M to the discounted value of taking that action. We do not describe here how such a function could be arrived at as there are a multitude of ways of estimating it, such as different schemas that may respond to different motivations. These values may depend on the value received from a downstream branch, so the robot cannot depend on getting all the motivation rewards. Figure 3 shows a road sign in Sicily that shows the way to two towns at opposite ends of the island. At some point, the driver will turn to head towards one or the other. In a maze, at some point, turning left might lead to cheese and turning right might lead to milk. If the robot rat is more hungry than thirsty, it will turn left, otherwise right. It will get either cheese or milk. Upstream of that bifurcation, an action leading that way will report positive values for both thirst and hunger motivations.

In the current state σ, the function $actionValue$ gives a function that maps motivations and actions to the expected value.

(a) Roadsign indicates the direction to two towns

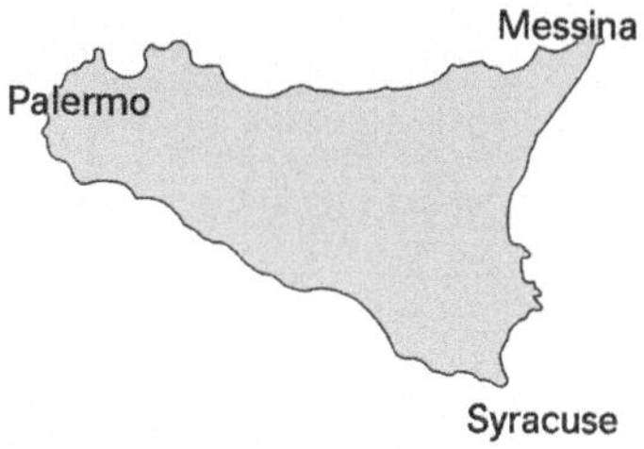

(b) Palermo is not in the same direction as Messina

Fig. 3. Route leads to two towns, but you will have to choose

$$\text{actionValue} : \epsilon \in \text{States} \mapsto ((\mu, \eta) \in M \times \text{Steps} \mapsto \mathbb{R}) \tag{3}$$

where M is the set of all motivations, $States$ is the set of possible states, and $Steps$ is the set of possible actions that the robot can take.

The best action to take is given by the loadedValue function, which calculates the fully loaded value of an action η for a specific motivation μ in a given state σ. If it is greater than zero, it may be chosen as the next step of the robot.

$$\text{loadedValue} : (\sigma, \mu, \eta) = actionValue(\sigma)(\mu, \eta)mRate(\mu) - \theta_{current} - i_{current} \tag{4}$$

The list of best actions is given by Eq. 5. If the list of actions is empty, there will be no change in action and the robot will do nothing after the completion of the current action.

$$bestActions = \underset{(\mu,\eta)\in(M,Steps)}{\arg\max}\ loadedValue(\sigma_{current}, \mu, \eta) \tag{5}$$

Equation 6 is None if there is no action that has a fully loaded value greater than zero.

$$chosenAction = \begin{cases} None & \text{if } empty(bestActions) \\ None & \text{if } loadedValue(\sigma_{current}, \\ & \quad bestActions_1[1], \\ & \quad bestActions_1[2]) < 0 \\ bestActions_R[2] & otherwise \end{cases} \tag{6}$$

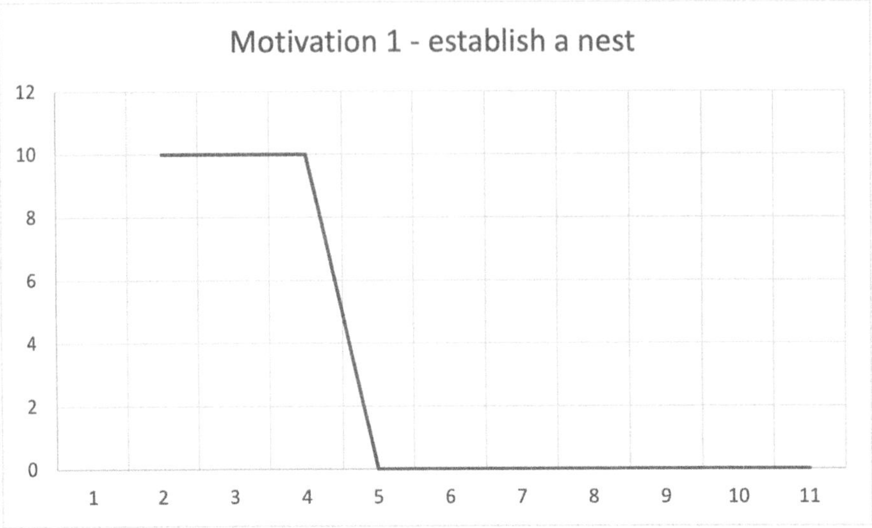

Fig. 4. A strong motivation to establish a home base

Motivation 1 Establishing a Home

Establishing a Home. In our experimental testbed, robots have an innate motivation to establish a home. They are not expected to build a nest, but they need to find an appropriate place to call home and register it on their cognitive map. We do not discuss the cognitive map in this paper, because our focus is on intrinsic motivations. The reflex of establishing a home involves finding a color patch. This motivation is very strong, but once the color patch has been found and the nest has been established, the motivation drops to zero, as shown in Fig. 4. Home is important in this experiment because the robots navigation capabilities are inexact due to wheel slippage. Their strategy for building good maps is to return to a known place frequently to improve accuracy of how they may their terrain and what they find within it.

Motivation 2 Return to the Nest Periodically. A motivation to return to the nest on a rhythmic basis is controlled by a natural extrinsic motivation whose strength follows a rhythm. An example of the periodicity of motivation strength is shown in Fig. 5. After a certain time, dictated by the rhythm of this motivation, the strength of the return-home motivation becomes too strong to resist, and the intended behavior occurs. The innate method of returning home involves going back in a direction that it believes leads home and then moving along the boundary until it finds the color patch. The home base is next to that. This is ensured by the way that a suitable place for the home is determined. It should be noted that learning can improve the ability to return home. We do not describe how this learning takes place in this article. The reader is referred

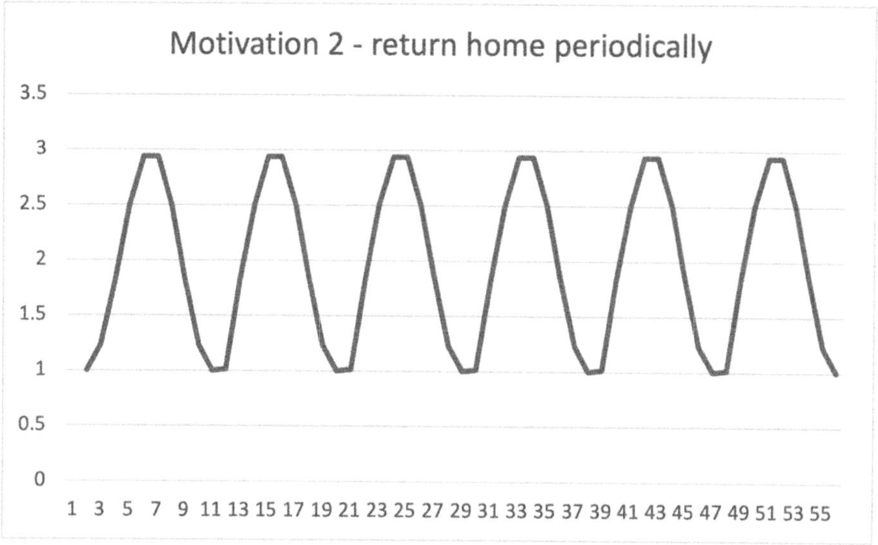

Fig. 5. A periodic motivation to return to the home base.

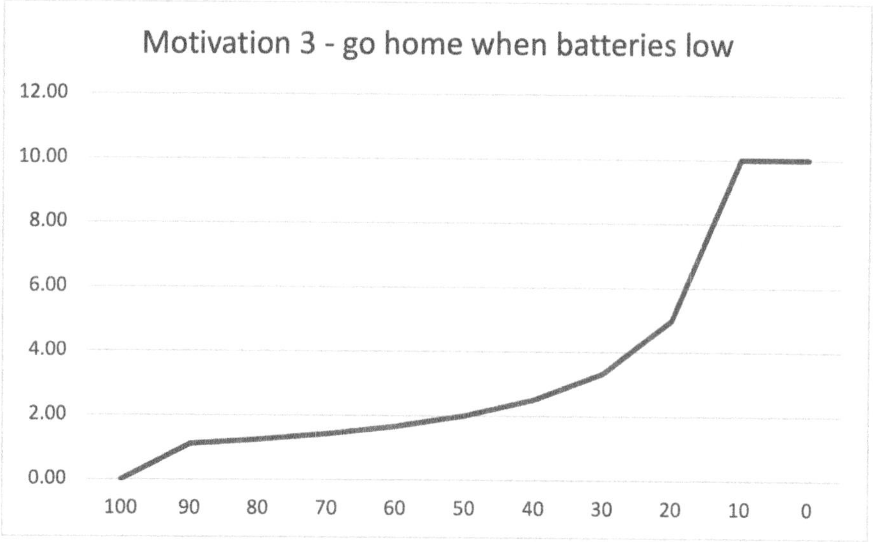

Fig. 6. As the battery level decreases the motivation to go home increases

to the article entitled "State-of-the-art Schema Mechanisms for Developmental Artificial Intelligence" in this collection.

Motivation 3 Return to the Nest When the Batteries Are Low. The robots should always return to their nest before their batteries are discharged.

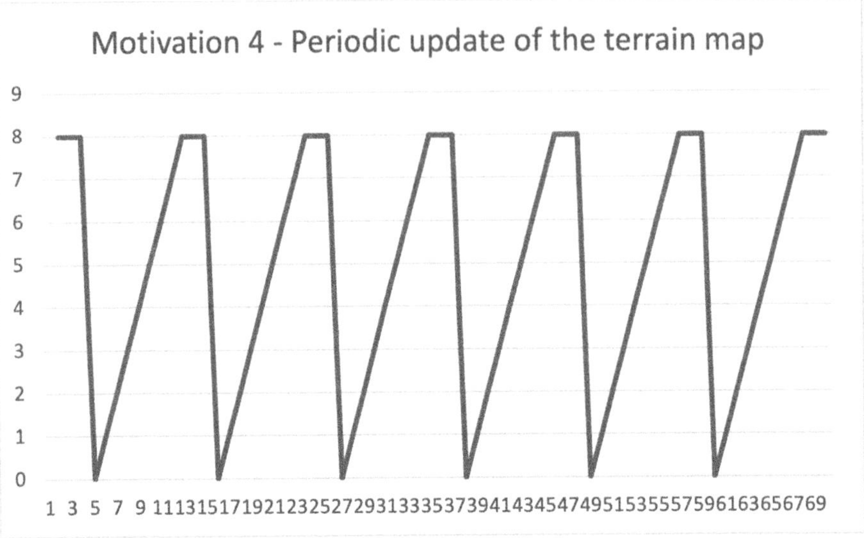

Fig. 7. "M4"

The valence of this behavior in relation to the level of battery charge is clear in Fig. 6. In this experiment, there is no charger because the robots cannot self-charge. We have another testbed that uses different robots, and there we were able to have the robot learn how to get back to the charger and locate itself on the charging panel for wireless charging and then continue to monitor its state of charge until it is ready to continue exploring its terrain. Returning to the charger is a skill that must be learned and practiced (Fig. 7).

Motivation 4 Maintain a Map of the Terrain. The terrain of a robot is delimited by a black line. Robots are motivated to have a recent map of their terrain. At the beginning, there is nothing written in its "allocentric" memory, so the motivation starts very high, but less than the maximum motivation level. All hexagons on the allocentric map are part of this calculation. Motivations three and four establish a rhythm of exploration.

Flocking. The tendency to follow other robots of the same type is controlled by the flocking settings.

If the separation force F_{sep} is large, there is a tendency for the robots to distance themselves from other robots; if F_{sep} is small, the cohesion force F_coh serves to decide whether to join a group or not. There is no separate period for flocking, but it is invoked as sleep time approaches, so the robots sleep close together in a pack formation. It is also during play; see Sect. 6.6 (Fig. 8).

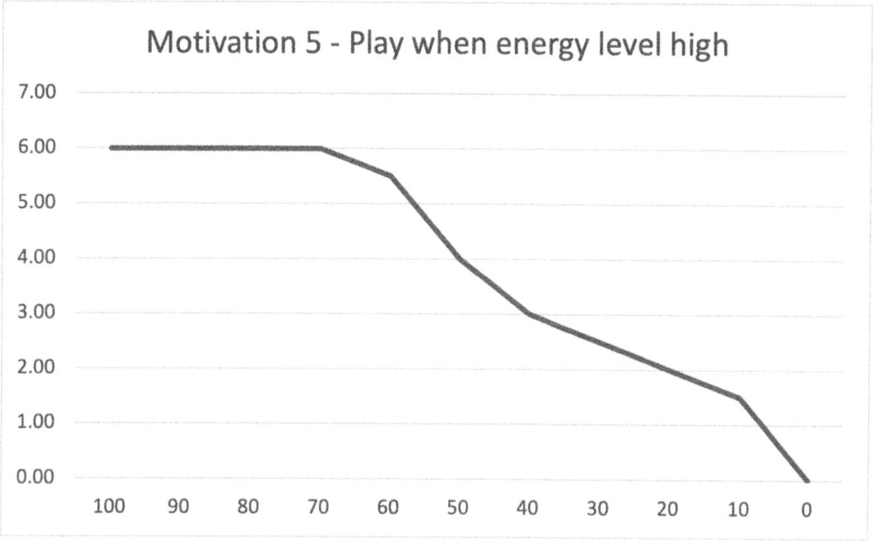

Fig. 8. "M5"

Motivation 5 Play with Other Robots. Simple play allows the robot to learn to increase curiosity and to increase its interest in being in the company of other robots. The key idea here is that when the robot has a lot of energy and nothing else has a higher need, the robot allows itself to socialize and explore. This helps the robot to put itself in a situation that is likely to lead to learning social skills and collaborative behavior as they stumble on things that can be done together (Fig. 9).

6.7 Remove Objects from the Terrain

Motivation 6 Remove Objects Found in the Terrain. The robots are motivated to remove objects they find in the terrain pushing beyond the boundary of the terrain. With these intrinsic motivations, robots can already learn simple collaboration, and by adding sounds and the recognition of sounds, they can learn to communicate with each other to guide their learned collaboration skills.

7 Case Study

To demonstrate how intrinsic motivation can be employed in a schema-based constructivist learning framework, we look at work by Georgeon, et al. [13] which introduces a model to simulate early mechanisms of emergent cognition based on enactive cognition and constructivist epistemology. The model, named the intrinsically motivated schema mechanism, aims to address three intertwined

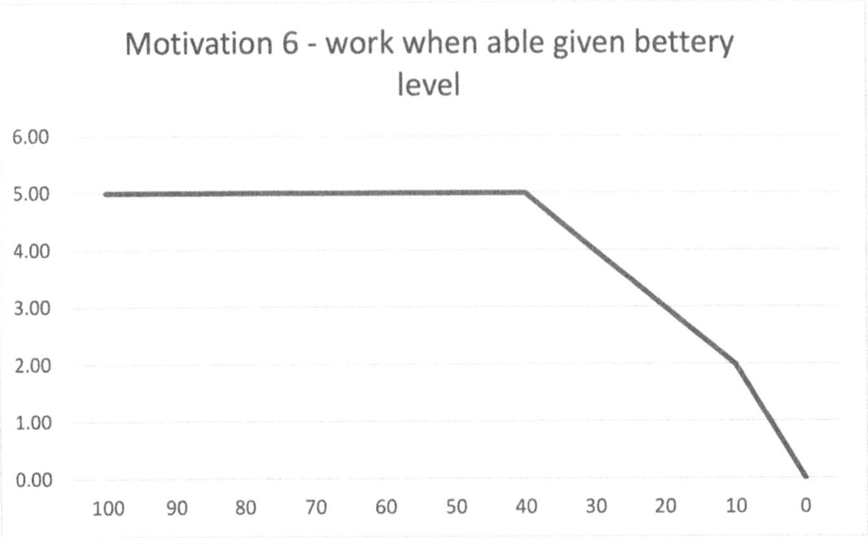

Fig. 9. "M6"

issues: intrinsic motivation, autonomously constructed internal state, and adaptive learning. The agent, referred to as Ernest, is driven by intrinsic motivations implemented as inborn proclivities [5,24] that guide its behavior proactively. Through interaction with its environment, Ernest autonomously learns regularities and hierarchical sequences of behaviors adapted to these regularities, developing a situational representation that generates expectations and activates adapted behaviors.

The model is based on three main mechanisms: reinforcement learning with intrinsic rewards, a hierarchical schema mechanism inspired by Piagetian constructivist epistemology, and episodic memory inspired by trace-based reasoning. Unlike traditional reinforcement learning, the agent does not receive rewards for reaching predefined goals but instead experiences intrinsic satisfaction from enacting behaviors. The agent's learning mechanism involves recording and reinforcing schemas based on interactions with the environment, with proclivity values representing intrinsic satisfaction.

Ernest's interactions are represented using schemas and interactions organized hierarchically. Primitive schemas are predefined, and higher-level composite schemas are learned through experience. The agent's scope, a data structure representing its current situation, is computed at the end of each decision cycle and used to select the next schema to enact. The scope includes interactions that have been completed and those whose schemas have passed a regularity sensibility threshold, ensuring that only sufficiently confirmed regularities are considered.

The algorithm follows two overlapping cyclical loops: the control loop, which involves selecting, enacting, learning, and assessing schemas, and the automatic loop, which enacts sub-schemas recursively. Schema selection is based on weighted propositions from activated schemas, with the highest summed proposition being selected. Schema enaction involves recursively enacting subinteractions and comparing feedback to expected statuses. If an expectation is not met, the enaction is interrupted, and the agent learns from the interactions actually performed and enacted.

The learning mechanism records higher-level schemas by associating context interactions with enacted interactions, with proclivity values set equal to the sum of sub-interactions' proclivity values. The scope assessment step ensures that the number of learned schemas remains manageable by including only interactions whose schemas have passed the regularity sensibility threshold.

The model was implemented in the Soar cognitive architecture, chosen for its flexibility and support for reinforcement learning mechanisms. Two experiments were conducted to illustrate the learning process and demonstrate the agent's emergent behavior. In the first experiment, Ernest learned to navigate a simple loop environment, increasing its average satisfaction over time. In the second experiment, Ernest was placed in a more complex environment and endowed with additional primitive schemas, allowing it to learn more elaborate perceptual behaviors.

The results show that Ernest can autonomously construct a representation of its situation and learn to use certain schemas to inform its perceptions and determine subsequent actions. The agent's behavior appears to exhibit early-stage cognitive phenomena, supporting the triple-autonomy hypothesis. However, the current algorithm faces limitations in managing a large number of learned schemas, finding spatial regularities, and avoiding local optima.

8 Conclusions

Intrinsic motivation mechanisms, as explored in this article, demonstrate significant potential in driving AI agents to learn, explore, and adapt in the absence of explicit goals. The approaches surveyed exhibit compelling behaviors in specific scenarios, offering valuable insights into how artificial agents can emulate aspects of human motivation. However, the challenge of designing a universal mechanism capable of operating effectively in all circumstances and addressing the full spectrum of inherent satisfactions remains unresolved.

One critical issue is ensuring that an intrinsically motivated agent does not fall into counterproductive states, such as idleness or excessive distraction. For example, an agent driven by curiosity may prioritize acquiring novel information or patterns at the expense of performing its primary tasks. In contrast, the absence of sufficient intrinsic motivation may result in the agent stopping all activity altogether. These challenges underscore the importance of balancing task execution and exploration, which may require dynamic adjustments to motivational parameters based on context and experience.

In addition, intrinsic motivation systems cannot be completely isolated from other considerations, such as task requirements, environmental constraints, and the level of expertise of the agent. The balance between learning and task execution is likely to evolve over the operational life of an agent, much as it does for humans. Young humans devote significant time to learning, while older humans often prioritize applying their knowledge. This shift is influenced not only by accumulated experience, but also by changes in motivation over time.

In biological systems, motivation is intricately linked to attention, physiological states, and emotions. Hunger, for example, drives people to seek and consume food, while certain innate drives, such as the motivation to find a partner, emerge at specific stages in life. Emotional states also profoundly influence motivation; a depressed individual may lack the drive to act, while a positive emotional state often enhances engagement and effort. These complex interactions suggest that any effective intrinsic motivation mechanism for AI agents must incorporate similar contextual and adaptive elements.

Given this complexity, it seems improbable that a single, straightforward intrinsic motivation mechanism will meet all the demands of general AI. The solutions discussed in this paper represent an important part of the puzzle, but they are not the complete answer. Intrinsic motivation must work in concert with other systems, including artificial emotions, task-specific priorities, and contextual awareness, to enable truly adaptable and autonomous agents.

Ultimately, we envision intrinsic motivation as deeply intertwined with artificial emotions, serving as an integrated framework for guiding agent behavior. Emotions could act as adaptive signals, helping agents decide whether to act or pause, seek knowledge or complete a task, engage socially or withdraw, embrace novelty, or find comfort in familiarity. Drawing from insights into human and animal behavior, we can move closer to designing AI systems that are not only effective but also relatable and capable of thriving in dynamic, complex environments.

Intrinsic motivation offers a promising foundation for creating self-directed adaptable AI agents. However, its full potential lies in its integration with broader systems that replicate the richness of human and animal decision making. Continued research is essential to bridge the gaps and unlock the transformative possibilities of intrinsic motivation in general AI.

Disclosure of Interests. The authors have no competing interests.

References

1. Andres, A., Villar-Rodriguez, E., Del Ser, J.: Towards improving exploration in self-imitation learning using intrinsic motivation. arXiv preprint arXiv:2211.16838 (2022)
2. Aubret, A., Matignon, L., Hassas, S.: An information-theoretic perspective on intrinsic motivation in reinforcement learning: a survey. arXiv preprint arXiv:2209.08890 (2022)
3. Baldassarre, G.: Intrinsic motivations and open-ended learning. arXiv preprint arXiv:1912.13263 (2019)

4. Bayer, H.M., Glimcher, P.W.: Midbrain dopamine neurons encode a quantitative reward prediction error signal. Neuron **47**(1), 129–141 (2005). https://doi.org/10.1016/j.neuron.2005.05.020
5. Blank, D.S., Kumar, D., Meeden, L., Marshall, J.: Bringing up robot: fundamental mechanisms for creating a self-motivated, self-organizing architecture. Cybern. Syst. **32**(2), 125–150 (2005)
6. Chapagain, D., Virányi, Z., Huber, L., Range, F., Schneider, L.A.: Aging of attentiveness in border collies and other pet dog breeds: the protective benefits of lifelong training. Front. Aging Neurosci. **12**, 73 (2020). https://doi.org/10.3389/fnagi.2020.00073
7. Chapagain, D., Virányi, L., Wallis, M., Müller, C.A., Range, F.: Aging of attentiveness in border collies and other pet dog breeds: the role of lifespan and age relative to life expectancy. Front. Psychol. **9**, 2007 (2018). https://doi.org/10.3389/fpsyg.2018.02007
8. Chowdhury, R., Guitart-Masip, L., Lambert, R., Dolan, J., Dayan, P., Düzel, R.: Dopamine restores reward prediction errors in older adults. Nat. Neurosci. **16**, 648–653 (2013). https://doi.org/10.1038/nn.3364
9. Deci, E.L., Ryan, R.M.: Intrinsic Motivation and Self-Determination in Human Behavior. Springer, Boston (1985). https://doi.org/10.1007/978-1-4899-2271-7
10. Deci, E.L., Ryan, R.M.: Self-determination theory and the facilitation of intrinsic motivation, social development, and well-being. Am. Psychol. **55**(1), 68–78 (2000). https://doi.org/10.1037/0003-066X.55.1.68
11. Di Lorenzo, F., Motta, C., Bonnì, S., et al.: Re-emergence of primitive reflexes as a sign of neurodegeneration in dementia. J. Neurol. **264**(1), 49–56 (2017). https://doi.org/10.1007/s00415-016-8315-6
12. Friston, K., FitzGerald, T., Rigoli, F., Schwartenbeck, P., Pezzulo, G.: Active inference: a process theory. Neural Comput. **29**(1), 1–49 (2017)
13. Georgeon, O.L., Ritter, F.E.: An intrinsically-motivated schema mechanism to model and simulate emergent cognition. Cogn. Syst. Res. **15**, 73–92 (2012)
14. Haber, N., Mrowca, D., Fei-Fei, L., Yamins, D.L.: Learning to play with intrinsically-motivated self-aware agents. arXiv preprint arXiv:1802.07442 (2018)
15. Kleim, J.A., Jones, T.A.: Principles of experience-dependent neural plasticity: implications for rehabilitation after brain damage. J. Speech Lang. Hear. Res. **45**(2), 373–391 (2002). https://doi.org/10.1044/1092-4388(2002/030)
16. Latyshev, A., Panov, A.I.: Intrinsic motivation in model-based reinforcement learning: a brief review. arXiv preprint arXiv:2301.10067 (2023)
17. Lawler, E.L., Lenstra, J.K., Rinnooy Kan, A.H., Shmoys, D.B.: The Traveling Salesman Problem: A Guided Tour of Combinatorial Optimization. Wiley, Chichester (1985)
18. Lehman, J., Stanley, K.O.: Abandoning objectives: evolution through the search for novelty alone. Evol. Comput. **19**(2), 189–223 (2011)
19. Li, S., Gureckis, T.M.: Aging and the role of curiosity in learning. Cogn. Affect. Behav. Neurosci. **18**, 1326–1344 (2018). https://doi.org/10.3758/s13415-018-0638-7
20. Meyerson, E., Miikkulainen, R.: Discovering evolutionary stepping stones through behavior domination. In: Proceedings of the Genetic and Evolutionary Computation Conference, pp. 49–56 (2016)
21. Milgram, N.W., Head, K., Zicker, D.A.: Cognitive enrichment in aging dogs: lessons from the dog brain. Neurobiol. Aging **23**(5), 737–745 (2002). https://doi.org/10.1016/S0197-4580(02)00020-5

22. Montague, P.R., Dayan, P., Sejnowski, T.J.: A framework for mesencephalic dopamine systems based on predictive hebbian learning. J. Neurosci. **16**(5), 1936–1947 (1996). https://doi.org/10.1523/JNEUROSCI.16-05-01936.1996
23. Nagashima, K., Morita, J., Takeuchi, Y.: Intrinsic motivation in cognitive architecture: intellectual curiosity originated from pattern discovery. Front. Artif. Intell. (2024)
24. Oudeyer, P.Y., Kaplan, F.: Intrinsic motivation systems for autonomous mental development. IEEE Trans. Evol. Comput. **11**(2), 265–286 (2007)
25. Park, D.C., Huang, C.M.: Neuroplasticity: changes in brain structure and function in aging. Nat. Rev. Neurosci. **14**, 87–97 (2013). https://doi.org/10.1038/nrn3399
26. Reynolds, C.W.: Flocks, herds, and schools: a distributed behavioral model. In: Proceedings of the 14th Annual Conference on Computer Graphics and Interactive Techniques (SIGGRAPH 1987), pp. 25–34. ACM (1987). https://doi.org/10.1145/37401.37406
27. Ryan, R.M., Deci, E.L.: Self-determination theory and the facilitation of intrinsic motivation, social development, and well-being. Am. Psychol. **55**(1), 68–78 (2000). https://doi.org/10.1037/0003-066X.55.1.68
28. Ryan, R.M., Deci, E.L.: Self-determination theory: Basic psychological needs in motivation, development, and wellness. Guilford Publications (2017). https://doi.org/10.1521/978.14625/28806
29. Samuel, A.L.: Some studies in machine learning using the game of checkers. IBM J. Res. Dev. **3**(3), 210–229 (1959). https://doi.org/10.1147/rd.33.0210
30. Schmidhuber, J.: Formal theory of creativity, fun, and intrinsic motivation (1990–2010). IEEE Trans. Auton. Ment. Dev. **2**(3), 230–247 (2010). https://doi.org/10.1109/TAMD.2010.2056368
31. Schultz, W., Dayan, P., Montague, P.R.: A neural substrate of prediction and reward. Science **275**(5306), 1593–1599 (1997). https://doi.org/10.1126/science.275.5306.1593
32. Stanley, K.O., Lehman, J.: Why greatness cannot be planned: the myth of the objective. Springer (2015)
33. Steels, L.: The autotelic principle. In: Iida, F., Pfeifer, R., Steels, L., Kuniyoshi, Y. (eds.) Embodied Artificial Intelligence. LNCS (LNAI), vol. 3139, pp. 231–242. Springer, Heidelberg (2004). https://doi.org/10.1007/978-3-540-27833-7_17
34. Strickland, E.: AI agents play hide-and-seek: an OpenAI project demonstrated "emergent behavior" by AI players - [News]. IEEE Spectrum **56**(11), 6–7 (2019). https://doi.org/10.1109/MSPEC.2019.8889898
35. Vansteenkiste, M., Ryan, R.M., Deci, E.L.: Self-determination theory and the explanatory role of psychological needs in human well-being. Psychol. Psychother. Theory Res. Pract. **74**(1), 13–22 (2006). https://doi.org/10.1348/000711106X162063
36. Walker, M.P., Stickgold, R.: Sleep, memory, and plasticity. Annu. Rev. Psychol. **60**, 49–79 (2009). https://doi.org/10.1146/annurev.psych.60.110707.163654
37. van der Werf, Y.D., Witter, M.P., Groenewegen, H.J.: The return of reflexive behavior in neurodegenerative diseases: a neuroanatomical perspective. Neurosci. Biobehav. Rev. **34**(5), 905–916 (2010). https://doi.org/10.1016/j.neubiorev.2009.10.007
38. Zheng, X., Ma, X., Shen, C., Wang, C.: Constrained intrinsic motivation for reinforcement learning. arXiv preprint arXiv:2407.09247 (2024)

Intelligence Tests for AI

Robert Laddaga(✉) and Paul Robertson

Dynamic Object Language Labs, Lexington, MA 08544, USA
{rladdaga,paulr}@dollabs.com

Abstract. The primary objective of this paper is to elucidate the purpose, methodology, and implications of intelligence testing in the context of artificial intelligence (AI). The discussion covers intelligence testing for humans and animals. It delves into the various forms of human intelligence and emphasizes the importance of assessing multiple dimensions. Furthermore, the Turing test is examined in relation to the broader question of intelligence testing for AI.

Keywords: artificial intelligence · intelligence testing · Turing Test

1 Introduction

1.1 Why Test Intelligence

There are two primary reasons for testing intelligence. The first is the binary question of whether intelligence is present or not. The second and more prevalent reason is to find out how much intelligence, and of what kind, there is in a specific subject, and also to be able to quantify the relational standing of one subject among others, in terms of intelligence.

We almost never divise binary tests for humans, perhaps because we simply assume them all to be intelligent[1]. In addition, the ability to take an intelligence test, regardless of the outcome, seems to suggest at least a modicum of intelligence. However, we appear to devise binary tests for animals, such as demonstrations of tool use in Wink: [59]. Wink notes: "Among corvids, parrots, dolphins, New-World and Old-World monkeys, and especially the great apes, there are many species that excel in intelligence. Apparently, they can recognize a context, make plans, and find innovative solutions. Intelligent behavior can be observed in many instances: Many animals make and use tools to open nuts, shells, snails, or birds' eggs. Sticks and wires are used to probe and spear hidden prey. Great apes use twigs to fish ants or termites out of their mounds or to get honey."

Intelligence assessment appears to consist of two distinct stages. The first stage involves determining whether a system or organism exhibits intelligence,

[1] There are many binary cutoffs that depend partly or wholly on intelligence: competence to stand trial, cognitive disability, etc., but these binary decisions use a cutoff on an intelligence scale combined with possibly other factors.

P. Robertson and O. Georgeon (Eds.): IWSSL 2024, LNCS 15457, pp. 121–150, 2025.
https://doi.org/10.1007/978-3-031-96325-4_6

assessed through a defined measure. The second stage, contingent on the presence of intelligence, evaluates the degree of intelligence, often in comparison to human capabilities. A scale that is not centered on human benchmarks would provide a more objective framework for comparing intelligence between species and artificial systems. By adopting a broader conceptualization of intelligence, it may be possible to mitigate cultural biases that have historically influenced human intelligence assessments.

When Alan Turing proposed his seminal test, the notion that a machine could exhibit intelligent behavior comparable to that of a human was highly controversial. At the time of Turing's writing, no computational systems existed that could satisfy the criteria of his test. The initial reactions to Turing's proposal divided opinion between those who deemed machine intelligence an impossibility and those who believed it to be an achievable goal. Early advances in artificial intelligence (AI) generated enthusiasm, demonstrating that computers could extend beyond mere calculations to solve constrained problems, such as chess. For decades, the Turing test was regarded as a distant milestone; however, significant developments brought the test to prominence once more.

In 2014, the chatbot Eugene Goostman, which simulated a 13-year-old Ukrainian boy, reportedly convinced 33% of human judges of its humanity during a Turing test competition organized by the University of Reading [57]. This achievement was widely publicized as the first instance in which an AI passed the Turing test. A decade later, in 2024, researchers at the University of California, San Diego conducted an experiment with GPT-4, an advanced language model. The participants engaged in conversations with GPT-4 and 54% misidentified AI as human, suggesting that GPT-4 effectively passed the Turing test under the given conditions [23]. One can certainly object that there are a wide variety of conditions for conducting a Turing test, and some are more stringent than others. For example, knowledge of LLM failings might improve one's chances of spotting one. However, we believe that there is a better approach than making the Turing test more stringent.

We propose that it is now necessary to advance beyond the Turing test and develop a more comprehensive framework for measuring intelligence. This new framework should integrate assessments of both animals and artificial systems, thus avoiding the anthropocentric biases inherent in traditional intelligence measures and the Turing test. We advocate for the adoption of a multidimensional intelligence assessment model, capable of evolving as our understanding of intelligence deepens. Such a model would situate animals, humans, and artificial intelligence along various dimensions, offering broader applicability. This approach not only enhances the objectivity of intelligence assessments, but also expands their utility beyond current applications in therapeutic and selection contexts.

1.2 How to Test Intelligence

What sorts of entity might be considered intelligent? Attempting to assess the intelligence of the Encyclopedia Britannica would appear to be a foolish notion.

Similarly foolish is to assess the intelligence of the Google search engine. Nevertheless, people have attempted to assess the intelligence of LLM based chatbots, despite them being much more like an encyclopedia or a search engine than like a subject that could be considered intelligent.

In 1980 the American philosopher John Searle wrote about a Chinese room thought experiment in "Minds, Brains, and Programs" [43]. In the experiment, an English-only speaker is in a room manipulating Chinese language symbols on the face of the room, according to instructions that he receives from outside the room. To outside observers who understand Chinese, it appears that whoever or whatever is in the room also understands Chinese. Of course, the English speaker does not understand Chinese. The point Searle was trying to make was that computers, which are only capable of manipulating symbols, can never actually understand anything, and hence cannot be intelligent. We don't agree with Searle's main point, nevertheless it is painful to watch the world fall for the same Chinese room deception with LLMs.

LLMs are notorious for their limitations. It is nearly universally agreed that they cannot reason or plan and often have problems doing simple math. They can often appear to reason or plan when they produce some reasoning or a plan they have trained on, but a more careful analysis demonstrates their inability [24].

We believe that intelligence ratings and rankings should be reserved for autonomous agents. Some people might consider that limitation to rule out various kinds of programmed entities, whether programmed by people or biology, such as robots or insects. We are more agnostic on the subject, although we do believe that adaptability is an important hallmark of intelligence. Also important for both intelligence testing and intelligence is motivation. Many unmotivated students have had their IQ underestimated, and animal researchers often use obtaining food as motivation for performing on tests. A good discussion of intertwined motivation and intelligence in sensemaking in artificial agents can be found in [16].

In the course of this paper, we will consider several different methods for testing intelligence. In order to develop such tests, we will consider the history of intelligence testing in humans, as well other notable efforts such as the Turing test and tests devised by biologists.

2 Intelligence Testing in Humans

The first intelligence test is normally credited to the French psychologist Alfred Binet along with his partner Theodore Simon. The French Ministry of Education was trying to decide which children with learning disabilities could be accommodated in regular schools and which would need special schools. Bineet thought objective evidence was a better method than a physician's opinion. The scale consisted of thirty tasks of graduated difficulty, from responding to the examiner up to repeating seven random digits, and reasoning from evidence. The scale was used to determine the child's intellectual age. In 1916 it was first revised by Stanford University psychologist Lewis Terman, and became known as the Stanford-Binet intelligence test. The test measured five factors including knowledge, quantitative reasoning, visual-spatial processing, working memory, and fluid reasoning. The term intelligence quotient (IQ) was proposed by German psychologist William Stern by comparing a child's scored mental age to their biological age and giving a ratio. Terman adopted the idea and multiplied the ratio by 100 to give a simple number. The Stanford-binet test has undergone five revisions, including shifting from mental age ratio scale to primarily a scale based on deviation from the population norm. Terman was quite successful in getting schools and the US military to use regular intelligence testing as part of ability grouping and recruiting efforts.

Due to various objections to the Stanford-Binet approach to intelligence testing, numerous alternative tests have been developed and similarly refined, including the Wechsler Intelligence Scale for Children (WISC) and Wechsler Adult Intelligence Scale (WAIS) tests, the Kaufman Assessment Battery for Children (K-ABC), the Woodcock-Johnson Cognitive Ability tests, the Differential Ability Scales, and the Cognitive Assessment System (CAS).

Individually administered intelligence tests generally require the subject to provide verbal responses and manipulate objects, making them less dependent on reading, but less suitable to mass or classroom settings.

Many of the critics of intelligence testing have objected to the single-dimensional score and pointed out that subtest scores are more valuable in diagnosing learning abilities and issues, but there has remained significant societal pressure on providing an overall score. For example, Harvard University psychologist Howard Gardner suggested that there were seven distinct intelligences: linguistic, logical-mathematical, musical, bodily-kinesthetic, spatial, interpersonal and intrapersonal, and later added existential and naturalist intelligences. We discuss the types of intelligence in Sect. 6.

A key problem with applying human intelligence tests to an AI is that such tests are published, and hence potentially available during training, and seeing a test before taking the test skews the result.

2.1 The Composition and Scoring of Modern Intelligence Tests

The Stanford Binet Intelligence Test includes four sections: Short-term memory, quantitative reasoning, verbal reasoning, and abstract/visual reasoning. Each takes approximately half an hour.

The four sections of the Stanford Binet test are scored by the number of correct answers. These numbers, when added together, constitute the test taker's intelligence score. Each score has a mean of 100 and a standard deviation of 16.

The Wechsler Intelligence Scale for Adults (WAIS) includes scores that combine to an intelligence index. The scored indexes are VCI, PRI, WMI and PSI (verbal comprehension, perceptual reasoning, working memory, and processing speed). Each index has two or three subtests from ten main subtests including Vocabulary, Information, Similarities, (under VMI), Digit Span, Arithmetic, (under WMI), Block Design, Matrix Reasoning, Visual Puzzles, (under PRI), Coding, and Symbol Search (under PSI). There are also five additional subtests as follows: Comprehension, Letter-Number Sequencing, Figure Weights, Picture Completion, and Cancellation.

3 Defects in the Turing Test

The Turing test has certainly captured the imaginations of generations of computer scientists, despite its many weaknesses and failings. To begin with, it is a binary test, which makes it less interesting. Secondly, it is based on the premise of an AI fooling a human tester into thinking the AI is human. It would seem then that if we design an AI to pass the Turing test, we are designing it to pull the wool over human eyes (not auspicious!).

It is worth considering the famous Asimov "Three Laws of Robotics":

1. A robot will not harm a human.
2. A robot will not allow a human to come to harm (unless there is a conflict with the first law).
3. A robot will not allow itself to be harmed (unless there is a conflict with the first or second law).

We certainly do not know how to build intelligent entities that are constrained by engineering to obey such laws, but in an era when we contemplate the regulation of AI technology, we should consider attempts to regulate AI behavior. One plausible regulation might be something like An AI may never falsely answer the question: "Are you an AI?"

However, the Turing test becomes an interesting test of human intelligence, when you consider, if required to conduct the question asking for the test, what questions should you ask? What questions are most likely to tease out the capable but nonhuman respondent? What if the human respondent is neuro-divergent (how would that affect the test)?

Efforts to conduct serious Turing tests have primarily shown that people are quite easily fooled, a result already familiar from numerous movies and television shows about how con men (confidence men and women) trick their victims.

4 How to Test AI for Intelligence

4.1 Recent Attempts to Create Tests for AI

A very interesting previous paper on intelligence testing for humans and AI is [9]. In that paper François Chollet argues that many AI researchers and some psychologists have measured intelligence in terms of skills exhibited, at games or puzzles. He further argues against this view because it is too easy to train for specific skills, particularly for an AI. Instead, he suggests that a better measure for intelligence be based on: "Algorithmic Information Theory, describing intelligence as skill-acquisition efficiency and highlighting the concepts of scope, generalization difficulty, priors, and experience." The principles for measuring intelligence put forth in his paper are as follows.

- measure abilities, not skills
- measure abilities via batteries of tasks without prior knowledge of tasks by either the system or developer.
- have "explicit standards regarding reliability, validity, standardization, and freedom from bias"
- measurement should not focus on anthropomorphic analyses of kinds of intelligence
- crystallized abilities in humans, like reading and writing should not be tested, since if a system can learn skills, then it should be able to learn crystallized abilities.

The article goes on to suggest a benchmark: the ARC-AGI benchmark. The benchmark consists of training examples and an output task. An example training set is a rectangular grid with filled squares of one color. The examples demonstrate that spaces enclosed by the filled squares should be filled with squares of a different color. The test is to correctly fill with the second color the enclosed spaces of the presented task. An example test is shown in Fig. 1.

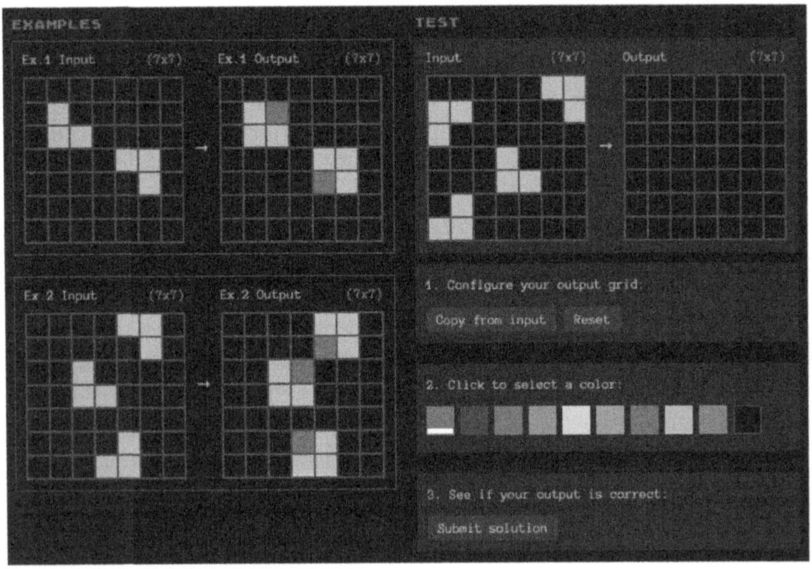

Fig. 1. Given the examples, identify the pattern and solve the test puzzle.

A much less interesting test set is being developed by contributors to "Humanitie's Last Exam", [36]. The authors note that LLM's are doing too well on current benchmarks, and believe that we should develop more difficult benchmarks of the same sort as we currently use, to better measure improved performance of LLMs. There is certainly nothing wrong with developing harder versions of the sort of tests LLMs are good at, so we can continue to improve LLMs, but there is a lot wrong with calling them by a dramatic name that mischaracterizes what LLMs do and what the test is good for.

4.2 Our View

One might have the view that intelligence itself is not general but specific to the kind of subject one characterizes as intelligent. That is not our view. We agree that intelligence is multifaceted, but that each facet can be more or less appreciated by any subject. The modes of testing and input-output methods may require variation, but the fundamentals of testing should hold.

Our view is that an important part of the core of intelligence testing should be based on stories, both story understanding and the elements of story construction. This is not an entirely new idea; among the most difficult questions in the original Binet test was one describing the following: "My neighbor has been receiving strange visitors. He has received in turn a doctor, a lawyer, and then a priest. What is taking place?" (Fancher, 1985).

Story understanding, constructing a story from parts, identifying missing parts of stories, and identifying lessons of stories provide excellent means of constructing valuable test questions.

5 The Other Side of Knowing It When You See It

Knowing it when you see it is a key part of the Turing test approach. By avoiding having a clear definition of intelligence, we simply use our judgment; the rules of the experiment are laid out in Turing's paper [55]. The other side is the knowing that something is not intelligent when we see it, for some unspecified but explainable reason. We have a growing list of such things. For example, we expect intelligent systems to learn from their mistakes. A system that continues to do the same thing that never worked in the past to achieve the same end is not intelligent, some would say. A system that never improves but simply has a fixed level of capability is not intelligent.

There are other examples of such non-intelligent indicators, such as not being able to make some level of chained reasoning, such as logical reasoning. A related indicator that has recently become prominent [24] is the inability to plan sequences of actions that lead to a solution. This approach is complicated for the same reason that the Turing test is powerful, because we need to be able to enumerate things that are necessary to be intelligent and we cannot necessarily enumerate them in advance.

Intelligence is often thought of as the ability of an entity to reason about the world and to act within that world for its better good, such as to avoid danger. In many cases, this involves manipulating objects in the world. If an AI has been trained on thousands of images from an annotated list of objects, it may be able to recognize one when it is presented as an image, but if it cannot interact with it, say a beach ball, it is not intelligent. This grounding problem is key to being intelligent. However, a person who can speak eloquently about beach balls and has learned from text what a beach ball is useful for and from training images what one looks like knows nothing about how to interact with one.

As outlined in the Introduction, a binary test determines whether the tested entity possesses or lacks the attribute we are assessing. Other tests, such as IQ tests, facilitate comparisons between entities, often involving individuals or populations. Individuals or groups strive to improve their test scores to gain advantages associated with higher scores. The principle of testing is aligned with the notion that "you get what you test for." Consequently, it is crucial to critically evaluate the nature of the test. Current LLM testing primarily assess LLMness, leading to competition among providers for higher scores and an increase in test difficulty. The most recent test of this type [36] has reached a point where only a select few people around the world can compete and benefit from its capabilities. This result has resulted in LLMs that excel in specific tasks but lack general-purpose capabilities that would be valuable to a broader audience. Instead of perpetuating narrowly focused tests, we propose developing a comprehensive test that encourages the development of AI with broader utility. To achieve this, we propose comparing our AI candidates with humans. Surprisingly, humans have been contemplating the nature of intelligence for centuries, driven by its practical application in identifying and addressing human deficiencies that can be improved through targeted education.

For this reason, in Sect. 6, we enumerate the dimensions of intelligence that have evolved. We will examine fifteen dimensions of intelligence. The reader will discern some patterns. Each dimension of intelligence is beneficial. That is, an individual who achieves a high score in that dimension will likely perform well in various aspects of life. In contrast, a low score warrants concern. For example, an individual with exceptionally low social intelligence (SI) will likely encounter difficulties. These dimensions exist, in part, because they are useful in designing curricula that assist individuals in improving their scores and thus mitigating their challenges. However, there are numerous deficiencies in the dimensions of intelligence that have been investigated to date. Many of them rely on self-reporting, which compromises the uniformity of test results. Furthermore, some dimensions lack such a reliable measure. There are challenges in distinguishing whether the test assesses intelligence or training. However, we believe that the answer lies, in part, in that body of work. The rationale for this belief is that dimensions have evolved from what is important to humans. Consequently, they are, by extension, what we aspire to find in our future robots. We would not desire a robot with a very low social intelligence score because they may be challenging to collaborate with. Similarly, we would not want our robot to possess a very low Moral Intelligence for obvious reasons. As the reader progresses through Sect. 6, it is beneficial to reflect on whether it would be detrimental if our future robot had a low score in that dimension.

Another pattern that the astute reader will notice is that there is significant overlap between some of the dimensions. Another is that each dimension has some utility in education, largely because we want to build curricula that target what is deemed to be important.

Not all aspects of an intelligence test are adequately covered by the existing literature on human intelligence. There are certain elements that we will need to incorporate which are specifically relevant to artificial systems.

6 Dimensions of Intelligence

The Turing test has served as an existence test for artificial intelligence and is binary, it is or it is not intelligent. For a long time, the test was seen as sufficiently difficult to achieve that it was a good test that avoided the pitfalls of trying to define what intelligence is. Today, however, the Turing test, as stated, is fairly easily passed, and a binary test is less useful.

It would be useful to be able to judge not just A.I. intelligence but also that of animal and humans. We would like to be able to ask not only if a creature or a manmade system is intelligent, but also how it compares to humans and animals, and other human-made artifacts.

We have frequently wanted to measure human intelligence and there has been much research on the topic. The word "Intelligence" suffers from being a convenient word that covers too much ground to be a useful scientific term. Someone who has poor spatial intelligence might be a genius at emotional intelligence. To be really useful, we would like to be able to measure the intelligence of AIs, humans, and animals against a set of measures.

Fortunately, there is a significant body of existing work that attempts to define the dimensions of intelligence. It may not be the definitive word on dimensions of intelligence, but it serves as a good starting point. It would be reasonable when applying dimensions to machines that some of the dimensions might need revision. It would not be reasonable to ignore the prior work. In this section, we summarize the previous work on multiple dimensions of intelligence.

Human intelligence has been categorized into various dimensions by researchers over time. These dimensions help explain how people process information, solve problems, and adapt to their environment.

It is useful to look at how we have chosen to measure intelligence along various dimensions of human intelligence and what purpose we put those measurement to, because the dimensions reflect what we have deemed important and AI will be judged along those same lines. A robot that fails to demonstrate emotional intelligence as we know it in humans will not be judged as intelligent. In fact, for AI to be accepted generally in society, these varieties of intelligence must be in place and score well compared to humans. Normally, scoring higher than a human would not be a problem, but scoring low could lead to rejection. For example, a robot with extraordinarily high social intelligence would be well received.

Whether we have done well or not in enumerating the kinds of intelligence is not our concern here. In this section, we explore not only the dimensions that have been proposed for measures of intelligence, but also how these intelligences are useful for a human and things that we do to reinforce them during the education of children or in people suffering with deficits.

This context allows us to better reflect on what measures we should be applying to AI.

1. General Intelligence (G-Factor). The G-factor or General Intelligence concept was introduced by British psychologist **Charles Spearman** [46] in the early 20th century. Spearman observed a consistent correlation between scores on different mental ability tests, suggesting a shared component that influences performance on tasks. The g-factor measures the overall cognitive ability of a person and consists of specific skills in various domains such as reasoning and problem solving. It represents a single underlying ability that influences performance across a variety of cognitive tasks. The theory suggests that individuals who perform well in one type of cognitive task (e.g., solving math problems) are likely to perform well in others (e.g., reasoning or vocabulary tasks), due to a shared cognitive capacity. The G-factor is viewed as the top level of a hierarchy, with more specific abilities (e.g., verbal, spatial, mathematical skills) branching out below it. Research in neuroscience has offered support for the G-factor in the form of functional characteristics of the brain, such as grey matter volume and processing speed.

As described in Sect. 2, general intelligence is typically assessed through **IQ tests**, such as the Wechsler Adult Intelligence Scale (WAIS), the Stanford-Binet Intelligence Scales, and the Raven Progressive Matrices, which is designed to isolate the G-factor by minimizing cultural and educational biases. These tests include subtests covering diverse domains (e.g., verbal comprehension, working memory, reasoning), and the overall score reflects the general intelligence factor.

John Carroll [7] proposed a hierarchical model in which the G-factorg is the broadest stratum at the top, specific abilities (e.g., fluid and crystallized intelligence) form the middle stratum, and narrow abilities occupy the lowest level. Raymond Cattell's Fluid and Crystallized Intelligence [8] differentiated general intelligence into two subtypes: Fluid Intelligence (Gf), which consists of problem-solving and adaptability, and Crystallized Intelligence (Gc), which refers to knowledge and skills gained through experience and education.

In practice, G-factor has been widely used, largely due to the convenience of having a single number, in education, where G correlates strongly with academic achievement, and in the workplace, where studies show a moderate to strong relationship between G and workplace productivity. Finally, G has been used to study cognitive decline in aging and neurodegenerative diseases.

Critics argue that intelligence is too multifaceted to be encapsulated by a single number or factor. Howard Gardner's [14] *Multiple Intelligences* theory, which is described below, and Robert Sternberg's later work *Triarchic Theory* [48] suggest that intelligence is domain-specific, rather than general. Some researchers argue that intelligence is better understood as a combination of domain-specific skills rather than a singular general ability.

2. Howard Gardner's Multiple Intelligences (MI). Howard Gardner introduced the **Theory of Multiple Intelligences (MI)** in his 1983 book *Frames of Mind: The Theory of Multiple Intelligences* [14,15]. This paradigm challenges traditional views of intelligence, which often emphasize a single measurable IQ score, by proposing that humans have a variety of distinct intelligences. Rather than proposing that intelligence is a function of the brain that grows with age and is somewhat fixed, MI views intelligence as not singular but consisting of various modalities, and those different intelligences may be valued differently across cultures and each intelligence develops at different rates and can be cultivated. Each of the proposed intelligences operates semi-independently, enabling people to learn, solve problems, and express creativity in diverse ways. Gardner's theory of multiple intelligences (1983) expanded the view of intelligence to include the following eight dimensions: 1) *Linguistic Intelligence*, the sensitivity to language, the ability to learn languages and use language effectively for communication and expression. This intelligence is often found in writers, poets, journalists, and lawyers, where language is used to create compelling stories or persuasive arguments. 2) *Logical-Mathematical Intelligence*, the ability to think, solve problems and solve problems, the aptitude for abstract reasoning and logical analysis, such as solving complex equations or designing algorithms. This intelligence is often found in scientists and mathematicians. 3) *Spatial Intelligence*, the ability to visualize and manipulate objects in 3-dimensional space. This intelligence is essential for navigators, architects, and artists. 4) *Musical Intelligence*, the sensitivity to sounds, tones, rhythms, pitch, melody, and harmony. This intelligence is involved in the composition of music and the recognition of tonal variations. 5) *Bodily-Kinesthetic Intelligence*, which includes control of body movements, manual dexterity, coordination of physical movement, and fine motor skills. This intelligence is especially found in athletes, dancers, and surgeons. 6) *Interpersonal Intelligence*, which refers to understanding and interacting effectively with others and involves skill in understanding and interacting with others. This intelligence is especially useful for teaching, giving therapy, leading or motivating a team and mediating conflicts. 7) *Intrapersonal Intelligence*, which refers to self-awareness and understanding of one's emotions and motivations, and 8)*Naturalistic Intelligence*, which refers to the sensitivity to nature, living organisms, and environmental patterns, Sensitivity to the natural world and an ability to recognize patterns in nature. This intelligence is seen in botanists, ecologists, and farmers.

Existential Intelligence. The concern for deep questions about existence, life, morality, purpose, and the universe has been proposed as a ninth intelligence and is believed to be found in spiritual leaders, philosophers, and theorists.

Practically, MI theory is leveraged in education to design instruction that engages students' diverse strengths, for example, using music to teach math concepts (musical intelligence) or outdoor activities to explain biology (naturalistic intelligence). MI theory helps people understand their strengths and align career choices with their dominant intelligences while encouraging individuals to explore and cultivate less dominant intelligences.

Critics [48] argue that Gardner's intelligences are better described as 'talents' or 'abilities' rather than distinct intelligences [58], and teachers can find it difficult to design curricula adapted to all intelligences in a classroom [3].

3. Emotional Intelligence (EI). Popularized by Daniel Goleman [18,18], EI involves the ability to perceive, understand, regulate and express emotions effectively. Emotional Intelligence (EI) refers to the ability to recognize, understand, manage, and influence one's own emotions and those of others. It was popularized by psychologist *Daniel Goleman* in the 1990s, but has roots in earlier work by researchers such as **Peter Salovey** and **John D. Mayer** [40].

Salovey & Mayer saw emotional intelligence as a cognitive ability with four branches: Perception of emotions, use of emotions to facilitate thinking, understanding emotions, and managing emotions. Petrides and Furnham [35] saw emotional intelligence as a set of personality traits such as empathy, assertiveness, and optimism.

Daniel Goleman's model identifies **five core components** of EI:

1) *Self-Awareness*, which involves recognizing and understanding one's emotions and their impact, such as recognizing and analyzing the feeling of anxiety before giving a presentation and preparing accordingly.
2) *Self-Regulation*, the management of emotions to adapt to circumstances and maintain control, for example, remaining calm during a heated discussion to effectively resolve conflicts.
3) *Motivation*, the use of emotions to drive achievement and persistence, which may be helpful when facing setbacks.
4) *Empathy*, understanding and sharing the feelings of others, such as sensing when a team member is stressed and offering support.
5) *Social Skills*, building and maintaining healthy relationships and the management of interpersonal dynamics such as may be necessary for team management.

In practice, EI has been linked, in the workplace, to leadership effectiveness, team collaboration, and conflict resolution, where it is claimed that high EI can foster a positive work culture by managing emotions and empathizing with employees. Social-emotional learning (SEL) programs integrated with IE in school curricula can improve students' academic success and social interactions. IE is related to better communication and empathy, strengthening personal and professional relationships. Finally, high EI helps individuals cope with stress, anxiety, and depression by fostering self-awareness and emotional regulation.

Emotional intelligence is measured by self-reporting, which some have criticized as not being reliable measures of actual emotional ability [30]. Others say that EI overlaps significantly with personality traits such as agreeability or extroversion. There is debate over how much EI contributes to success compared to cognitive intelligence (IQ) or personality. There is little evidence to support its usefulness.

4. Practical Intelligence. Proposed by Robert Sternberg [47,52] as part of the Triarchic Theory of Intelligence, practical intelligence refers to the ability to solve everyday problems, adapt to real-world situations, adapt to new environments, and effectively manage everyday tasks. The Triarchic Theory of Intelligence, also includes *analytical intelligence* and *creative intelligence.*

Sternberg argued that traditional IQ tests measure primarily analytical intelligence, neglecting practical intelligence, which is critical for success in life. Practical intelligence is sometimes referred to as "street smarts" because it involves applying knowledge to navigate and address life's challenges, rather than academic or theoretical problems.

Practical intelligence consists of three sub-parts: 1) The ability to adapt to or select environments that fit personal strengths and needs such as adjusting communication styles based on the audience. 2) Finding effective solutions to real-world challenges rather than hypothetical scenarios, such as troubleshooting technical issues or managing interpersonal conflicts. 3) Knowledge gained through experience such as understanding workplace dynamics without formal training and 4) The "how-to" aspects of problem-solving, emphasizing skills like decision-making and implementation.

Practical intelligence is often assessed indirectly because it lacks formal, standardized tests. Situational Judgment Tests (SJTs) are used to evaluate how individuals handle real-world scenarios, such as asking participants to choose the best course of action in a workplace conflict. Tacit Knowledge Inventory measures knowledge gained from practical experiences. See Steinberg's [52] studies on managerial tacit knowledge.

Practical intelligence is criticized as being difficult to quantify, leading to subjective or context-dependent evaluations [20,49]. Critics also argue that tacit knowledge may overlap with experience or personality traits and also that practical intelligence may not generalize across cultures where "practicality" is defined differently.

Despite its measurement challenges, practical intelligence appears to be essential for the workplace, where it is considered critical for leadership, teamwork, and navigation of organizational structures. Managers with high practical intelligence excel in decision making and conflict resolution. In schools, practical intelligence complements analytical skills, encouraging problem solving that applies to daily life, such as teaching students financial literacy for real-world application. Finally, entrepreneurs with strong practical intelligence identify opportunities and manage risks effectively.

5. Creative Intelligence. Creative intelligence, a key component of Sternberg's Triarchic Theory [14,47], refers to the ability to generate new ideas and think innovatively. It encompasses problem solving through unconventional approaches and is particularly valuable in fields that demand adaptability, creativity, and design thinking. Research in psychology, education, and organizational leadership has extensively examined the role of creative intelligence in human cognition and performance.

Howard Gardner's theory of multiple intelligences [14,15] underscores the diverse nature of cognitive abilities, with creative intelligence intersecting with interpersonal, linguistic, and spatial intelligences. This perspective highlights the multidimensional nature of human intelligence. In contrast, Sternberg's Triarchic Theory classifies intelligence into three distinct components: analytical, creative, and practical. According to Sternberg, creative intelligence is characterized by the capacity to navigate novel situations and devise original solutions beyond conventional paradigms.

Aspects of creative intelligence include *Fluency*: Generating numerous ideas or solutions, *Flexibility*: Adapting to different perspectives or methods, *Originality*: Producing unique and innovative results, and *Elaboration*: Developing and refining ideas into practical applications [2,39].

Practical uses of creative intelligence include encouraging students to think critically and solve problems creatively as a way of enhancing learning outcomes in schools. Activities such as brainstorming, storytelling, and arts integration foster creative intelligence. Business leaders with high creative intelligence can drive organizational innovation and adapt to rapidly changing markets. Creative intelligence plays a crucial role in creating user-centered designs and innovative technological solutions.

The following approaches are employed to enhance creative intelligence in humans: Engage in *divergent thinking* activities such as brainstorming or lateral thinking exercises, cultivate a *growth mindset* to embrace challenges and learn from failures, foster curiosity and explore diverse disciplines to expand the cognitive toolkit, and practice mindfulness and relaxation techniques to improve mental clarity and creativity.

6. Analytical Intelligence. The third component of Sternberg's Triarchic Theory [47], which includes logical reasoning and academic problem solving, is the cognitive capacity to analyze, evaluate, and solve well-defined problems systematically. Sternberg emphasized that analytical intelligence is necessary for tasks that involve evaluation and judgment, particularly in structured and formal settings.

Analytical intelligence is often linked to logical reasoning, critical thinking, and the ability to process and organize information. This form of intelligence is frequently associated with academic success and traditional measures of IQ. Among the capabilities that fall under analytical intelligence are 1) Breaking down complex problems into simpler components and finding solutions. 2) Using structured approaches to reach conclusions. 3) Evaluating arguments, identifying flaws and making reasoned judgments. 4) Identifying relationships and structures in abstract data and 5) Interpreting data, organizing them meaningfully, and making sound decisions [8,32,46].

Analytical intelligence is the most recognizable form of intelligence because it is akin to standardized IQ tests. It is measured using standardized IQ tests, such as the Wechsler Adult Intelligence Scale (WAIS) or Stanford-Binet, to assess analytical intelligence through problem solving and logical reasoning tasks. Educational assessments such as tests designed to evaluate critical thinking, comprehension, and reasoning skills are also used.

Analytical intelligence is found in academics, which emphasizes mathematics, science, and logic-based disciplines including artificial intelligence, and also in careers in engineering, law, finance, data analysis, and research, which require strong analytical skills. Decision-making in business, such as analyzing market trends or interpreting financial reports, also requires analytical intelligence.

Given the central role given to IQ testing and the perceived view that a high IQ is correlated to success, a lot of effort has been devoted to improving people's aptitude for this kind of thinking. Beyond mathematics, there are methods for practicing critical thinking, such as debates, discussions, and analysis of case studies. Certain types of games such as mastermind, chess, GO, and Sudoku are encouraged to develop problem solving abilities and analytical intelligence. Data analysis and learning to understand data by presenting them in the form of charts and graphs is seen as a practice that can enhance analytical thinking. Exploring new academic fields is seen as a way of broadening analytical capabilities.

7. Social Intelligence. Edward Thorndike [54] introduced the term Social Intelligence, which refers to the ability to navigate and manage social relationships and environments. Thorndike introduced social intelligence as distinct from academic and mechanical intelligence. He viewed it as practical intelligence for human interaction. Social intelligence (SI) is the ability to effectively navigate and manage social relationships, understand others' emotions and behaviors, and adapt to social environments. Edward Thorndike described it as the "ability to understand and manage men and women and to act wisely in human relations." Over time, it has been explored in psychology, education, and leadership contexts. Goleman emphasized, in his books *Emotional Intelligence* (1995) and *Social Intelligence* (2006) [18,19], that emotional competencies, such as empathy and self-regulation, and the interplay between emotional and social intelligence are effective in leadership and interpersonal success. In Gardner's *Theory of Multiple Intelligences* (1983) [14], interpersonal intelligence closely aligns with social intelligence. It involves understanding others' intentions, emotions, and motivations. Albrecht proposed in *Social Intelligence: The New Science of Success* (2006) [1], five key dimensions of social intelligence, summarized as "S.P.A.C.E." These five dimensions are as follows: 1) *Situational Awareness*, 2) *Presence*, 3) *Authenticity*, 4) *Clarity*, and 5) *Empathy*.

The skills that characterize social intelligence are the ability to understand and share the feelings of others (empathy), recognizing social cues, group dynamics, and cultural nuances (social awareness), building and maintaining positive relationships (interpersonal skills), modifying behavior to fit different social contexts (adaptability), managing disagreements effectively and fostering cooperation (conflict resolution) and understanding how to present oneself appropriately in social settings (self presentation).

Social intelligence is measured through observational methods, self-report questionnaires, and behavioral assessments, such as the George Washington Social Intelligence Test (GWSIT), which is one of the earliest tools for assessing social intelligence, and the Social Skills Inventory (SSI), which measures verbal and nonverbal social competence.

Social intelligence deals directly with our relationship with other intelligent entities, mostly humans, but could be easily extended to animals and perhaps soon to artificial intelligences. Effective leaders who use social intelligence to inspire, motivate and resolve conflicts within teams see it as demonstrated. Emotional connections with others improve trust and cooperation. Teachers use social intelligence to create inclusive classrooms and address diverse student needs. Social intelligence helps to understand consumer behavior, negotiate deals, and maintain professional relationships. Social intelligence is crucial to building strong friendships, romantic relationships, and family bonds. Whereas analytical intelligence is perceived as crucial for analytical endeavors, social intelligence is important for almost all human activities because we are a social species.

Given its almost universal importance and, perhaps surprisingly, that many people feel inadequate in social interactions, it is less surprising that there have evolved methods for improving this dimension. Social intelligence can be improved by practice, as can perhaps all measures for intelligence. Strategies for improving social intelligence include paying attention to verbal and nonverbal cues during conversations. Practicing seeing situations from the perspectives of others, engaging in genuine interactions and seeking common ground, learning techniques such as collaborative problem solving, and exposing oneself to various social settings to understand different norms of behavior.

As with other dimensions of intelligence that depend upon self-reporting methods as the primary form of measurement, critiques of Social Intelligence have cited difficulties in measuring it objectively. The overlap with emotional intelligence, leading to ambiguity in definitions and cultural variability in social norms, is also criticized. These problems make application challenging.

8. Cultural Intelligence (CQ). Earley and Ang (2003) [11] first introduced the term cultural intelligence. Cultural intelligence [27,34,53,56] is a concept rooted in the broader field of intelligence research, particularly inspired by multiple intelligences, emotional intelligence (EQ), and social intelligence. The term is based on theories from psychology, sociology, and organizational behavior. Cultural intelligence is defined as an individual's capability to function effectively in culturally diverse settings. It extends beyond cognitive intelligence (IQ) and emotional intelligence (EQ) by emphasizing adaptability and effectiveness in intercultural interactions. It is the ability to relate to and work effectively in culturally diverse situations. It encompasses knowledge, skills, and attitudes that enable individuals to navigate cultural differences effectively. Cultural intelligence is crucial in an increasingly globalized world where individuals interact with people of diverse cultural backgrounds in professional, social, and personal contexts. Beyond sensitivity, it goes beyond the ability to exhibit cultural sensitivity and reflectivity to include knowledge of other cultures.

Cultural intelligence is broken down into four main pillars as follows: 1) *Cognitive (Knowledge)* Understanding cultural norms, practices, and conventions. This includes knowing about different cultural systems (legal, economic, social) and cultural values. 2) *Meta-cognitive (Awareness)* Reflecting on and adjusting one's cultural knowledge and perceptions. This involves planning, monitoring, and revising cultural assumptions. 3) *Motivational (Drive)* The interest, confidence and persistence to adapt to culturally diverse settings. It is the emotional and motivational aspect of engaging with other cultures. 4) *Behavioral (Action)* The ability to exhibit culturally appropriate verbal and non-verbal behaviors. This includes flexibility in communication styles and adapting behaviors to suit different cultural contexts.

Cultural intelligence is increasingly diverse in the world. It helps prevent misunderstandings and fosters effective cross-cultural communication, and it encourages collaboration and mutual respect. Leaders with high cultural intelligence are better equipped to manage culturally diverse teams and organizations. Cultural intelligence facilitates the understanding and constructive resolution of cultural conflicts and helps individuals adapt to new cultural environments.

The theoretical foundation of CQ draws on Howard Gardner's Multiple Intelligences Theory [14] Specifically, interpersonal and intrapersonal intelligences relate to understanding and managing interactions with people of different cultural backgrounds. Sternberg's triarchic theory of intelligence [47] cultural intelligence aligns with Sternberg's practical intelligence, as it involves a real-world application of cultural awareness. Social Learning and Cross-Cultural Adaptation - The concept is influenced by Bandura's [4] (1977) social learning theory, which suggests that behavior and cognition are shaped through observation and social interactions.

Cultural intelligence is typically assessed using the Cultural Intelligence Scale (CQS), developed by Ang et al. (2007). This is a self-report questionnaire that evaluates the four components on a Likert scale. Other tools include the Intercultural Development Inventory (IDI), which measures intercultural competence and sensitivity, the Multicultural Personality Questionnaire (MPQ), which assesses traits related to successful adaptation in diverse environments, and the Global Competency Inventory (GCI), which evaluates global leadership skills, including cultural intelligence.

Proponents claim that 1) companies with culturally intelligent teams achieve better results in international markets. 2) It improves teaching and learning experiences in multicultural classrooms. 3) Cultural intelligence improves patient-provider relationships in diverse healthcare settings. and 4) Diplomats with high cultural intelligence navigate international relations more effectively.

However, critics of cultural intelligence raise concerns about its conceptual foundation, measurement, and real-world application. In particular, critics argue that cultural intelligence does not introduce fundamentally new ideas but instead repackages existing concepts of emotional intelligence, social intelligence, and intercultural competence. Cultural intelligence shares similarities with intercultural competence, global mindset, and adaptability, leading some scholars to question whether it is a genuinely distinct construct or simply a new label for well-established ideas. Some critics suggest that cultural intelligence may reinforce cultural essentialism, the idea that cultures have fixed characteristics, potentially leading to stereotyping rather than genuine intercultural adaptability. As with other measures that depend on self-reporting methods, the cultural intelligence scale (CQS), the most widely used tool to measure CQ, relies on self-reports, which may be prone to social desirability bias -individuals rating themselves higher than their actual abilities. Some researchers argue that cultural intelligence measures cultural adaptability in general rather than in specific cultural contexts, making it less predictive of real-world behavior in unique settings. Although cultural intelligence is associated with intercultural effectiveness, some critics argue that existing research does not strongly demonstrate that higher cultural intelligence directly leads to better outcomes in cross-cultural interactions. Although cultural intelligence training is widely promoted, some critics question whether short-term workshops can meaningfully enhance a person's ability to function in diverse cultural settings. Long-term exposure and genuine intercultural experiences may be more effective. Some scholars note that much of the CQ research originates from western corporate environments, raising concerns about ethnocentric biases in its application and whether it is equally relevant in non-Western or indigenous cultural frameworks. Critics argue that cultural intelligence focuses on individual adaptability while overlooking structural and organizational barriers that affect cross-cultural interactions, such as institutional biases and systemic discrimination. Some researchers criticize cultural intelligence for treating culture as a tool for personal or corporate gain, rather than fostering genuine intercultural understanding and mutual respect. Finally, there is concern that high-CQ people might use their cultural knowledge to manipulate rather than collaborate, particularly in negotiations, politics, or international business.

9. Fluid and Crystallized Intelligence. These two complementary components of general intelligence were introduced by the psychologist Raymond Cattell [8,21,22,31,41] in the mid-20th century. These concepts describe how individuals acquire, apply, and adapt their intellectual abilities in different contexts and stages of life.

1. Fluid intelligence (Gf) refers to the ability to think logically, solve new problems, and identify patterns without relying on prior knowledge or experience.

It involves abstract thinking, reasoning, and problem solving and is independent of acquired knowledge, such as vocabulary or cultural influences. It peaks during adolescence and gradually declines with age.

Examples of fluid intelligence include solving puzzles, identifying logical patterns, and strategizing in new or unfamiliar situations.

It relies on working memory and cognitive flexibility and is associated with neural efficiency and the brain's prefrontal cortex.

2. Crystallized Intelligence (Gc)

Crystallized intelligence is the accumulation of knowledge and skills acquired through education, experience, and cultural exposure. It involves learned information, such as vocabulary, general knowledge, and expertise in specific domains. It improves with age as people gain more experience and education. It remains stable or even increases well into old age.

Examples of crystallized intelligence include solving a crossword puzzle using known words, recalling historical facts, and applying mathematical formulas learned at school.

Crystallized intelligence draws on long-term memory and semantic knowledge and is associated with the temporal and parietal lobes of the brain.

Fluid intelligence helps to acquire new knowledge that contributes to the development of crystallized intelligence. Although fluid intelligence declines with age, crystallized intelligence tends to remain stable or improve, compensating for cognitive decline in older adults.

Fluid and crystallized intelligence are measured using the Wechsler Adult Intelligence Scale, and Raven's Progressive Matrices assess aspects of both types of intelligence.

The practical value of these two measures of intelligence include tailoring teaching methods to enhance both problem solving skills (Gf) and knowledge acquisition (Gc) and, in the workplace, leveraging fluid intelligence for innovation and crystallized intelligence for expertise-based tasks. For elder care, it is used to design interventions to maintain fluid intelligence and optimize crystallized knowledge in older adults.

10. Existential intelligence refers to the ability to ponder deep questions about human existence and is one of the multiple intelligences proposed by Howard Gardner [14] in his *Theory of Multiple Intelligences.* It involves sensitivity to deep questions about existence, life, and death, including the meaning of life, why we die, and what our place is in the universe. It also involves sensitivity and the ability to tackle questions about reality, purpose, and existential dilemmas. People with high existential intelligence are often reflective, philosophical, and interested in life's "big questions."

The four facets of existential intelligence can be characterized as reflectiveness, philosophical inquiry, spiritual awareness, and ethical sensitivity. Respectively these entail For *Reflectiveness*, a natural tendency to reflect on abstract concepts like the nature of being, mortality, and consciousness. For *Philosophical Inquiry*, an interest in exploring philosophical ideas and participating in discussions about the purpose of existence. For *Spiritual Awareness*, often associated with increased awareness of spiritual and metaphysical issues. And for *Ethical Sensitivity*: A strong sense of moral responsibility and concern for the well-being of humanity.

High levels of existential intelligence are often found in philosophers, theologians, and spiritual leaders, as well as artists and writers exploring existential themes and individuals who ask profound questions and seek understanding of life's ultimate meaning.

Howard Gardner [14] introduced the idea of existential intelligence in his book *"Intelligence Reframed: Multiple Intelligences for the 21st Century"* (1999) [15]. He described it as "the intelligence of big questions" and debated its inclusion in the list of intelligences because of challenges in measuring it empirically. Sternberg [51] in *The Evolution of Intelligence: Are Humans the Only Creatures Who Ask 'Why?'* explores the evolutionary basis for existential thought and its relation to human intelligence. Armstrong's 2009 paper [3] *Multiple Intelligences in the Classroom.* provides strategies for educators to nurture various intelligences, including existential intelligence, in students. *Emmons' 1999 paper* [12] *The Psychology of Ultimate Concerns: Motivation and Spirituality in Personality* explores the connections between existential questions and human motivation, spirituality, and personality development.

In Education, teachers encourage students to engage philosophical and ethical questions to foster deeper critical thinking. Existential intelligence is used in *therapy and counseling* to address existential questions that can help people find meaning and purpose in life. Leaders with high existential intelligence often inspire others by focusing on shared human values and higher goals. Works that explore existential themes resonate with people due to their universal relevance.

11. Moral Intelligence. *Moral Intelligence* refers to the ability to understand right from wrong and to behave according to ethical principles. It encompasses the ability to recognize moral issues, make ethical decisions, and act according to core values such as honesty, responsibility, compassion, and respect for others.

This concept is rooted in the fields of psychology, philosophy and education and is considered essential for personal integrity and social harmony. Moral intelligence encompasses three concepts: *Knowing*, the capacity to recognize and understand moral principles. *Feeling*, the capacity to develop empathy and emotional alignment with ethical values, and *Acting*, the demonstration of behaviors consistent with moral beliefs. Integrates cognitive, emotional, and social intelligence to guide ethical behavior.

While not being part of Gardner's multiple intelligence, moral intelligence supports the idea that different types of intelligence, by building upon how interpersonal and intra-personal intelligence contributes to moral reasoning. Lawrence Kohlberg's 1984 paper [25] *Essays on moral development* outlined six stages of moral development, ranging from obedience to avoid punishment (preconventional) to adherence to universal ethical principles (post-conventional). Carol Gilligan's Ethics of Care [17] emphasized the role of care and empathy in moral reasoning, particularly highlighting differences in how genders approach ethical dilemmas. Lennick and Kiel [26], dissect moral intelligence as the following four components: *Integrity*: acting consistently with one's moral values. *Responsibility*: taking ownership of one's actions and their consequences. *Compassion*: demonstrating empathy and care for others and *Forgiveness*: Letting go of resentments and fostering reconciliation. Kohlberg's book entitled *Essays on Moral Development, Volume 1: The Philosophy of Moral Development* [25] is a foundational text on moral reasoning and stages of moral development. In

Goleman's [18] book *Emotional Intelligence: Why It Can Matter More Than IQ* he highlights the role of empathy and ethical decision-making.

Moral intelligence is important because humans have been successful in forming societies that live and work together under a shared set of moral virtues; it is the glue that holds us together. People who lack moral intelligence cannot function in our society. It is noted that leaders with moral intelligence foster trust and ethical organizational cultures. It helps in mediating disputes and promoting fairness, while encouraging respect and compassion in diverse societies. It guides people to make ethical choices that align with their values. For example, it leads us not to cheat even when no one is watching, to make business decisions that prioritize people and the environment over profit, and to stand up against injustice.

Moral intelligence, in practice, offers important value and is used in education, corporate ethics, parenting, and healthcare. In education, it integrates moral reasoning and character education by informing curricula development to teach students about fairness, integrity, and empathy. Borba's book [5] *Building Moral Intelligence: The Seven Essential Virtues that Teach Kids to Do the Right Thing* is a practical guide for parents and educators to nurture moral intelligence in children. In corporate ethics, moral intelligence serves to promote ethical decision making and leadership in businesses. The book *Moral Intelligence: Enhancing Business Performance and Leadership Success* by Lennick and Kiel [26] focuses on how moral intelligence underpins effective leadership and organizational success. Gilligan's book [17] *In a Different Voice: Psychological Theory and Women's Development.* explores the ethics of care and the role of empathy in moral intelligence.

12. Kinesthetic Intelligence refers to the ability to use physical movements and manipulate objects skillfully, often tied to sports or fine motor skills, and the ability to use one's body effectively to solve problems, create products, or express ideas and emotions. *Kinesthetic Intelligence*, also known as *Bodily-Kinesthetic Intelligence*, is one of the intelligences identified by Howard Gardner [14] in his *Theory of Multiple Intelligences.* Individuals with high kinesthetic intelligence excel in physical activities and have a higher sense of body awareness. Gardner argued that intelligence is not a singular ability, but a combination of multiple modalities, of which bodily-kinesthetic intelligence is a key component. Shapiro's book [44] *Embodied Cognition* discusses how bodily movements and actions contribute to cognitive processes, closely related to kinesthetic intelligence.

Kinesthetic intelligence consists of four key components, as follows: *Body Awareness*, the sensitivity to how one's body moves and interacts with the physical environment. *Coordination*, hand-eye coordination and motor skills. *Physical Expressiveness*, the ability to communicate emotions and ideas through bodily movements and *Learning through Action* a preference for hands-on activities and learning by doing.

Athletes use their bodies to perform complex physical feats, dancers express ideas and emotions through choreography and movement, surgeons perform precise physical tasks that require fine motor skills, and sculptors or carpenters create objects using their hands.

Although, as with most dimensions of intelligence, individuals may differ in their innate aptitude, they can also improve their capabilities through targeted practice. Regular participation in sports, dance or physical tasks that require coordination and skill, focusing on hands-on projects and activities, participating in activities that develop body awareness and control, such as yoga or martial arts, and exploring crafts, drama, and other forms of bodily expression, all contribute to improved kinesthetic intelligence.

In education, hands-on activities, such as experiments, role-playing, and physical games, are used to engage students with strong kinesthetic intelligence. Other strategies, such as incorporating movement-based learning strategies, such as building models or acting out scenarios, similarly help develop fine motor skills and coordination. Armstrong's book [3] *Multiple Intelligences in the Classroom.* explores strategies to foster different types of intelligence, including kinesthetic intelligence, in educational settings.

In sports, athletes and performers develop skills through targeted practice and body awareness exercises and a focus on improving motor coordination and physical efficiency. Csikszentmihalyi's book [10] *Flow: The Psychology of Optimal Experience.* explores the concept of "flow," often experienced by individuals engaged in kinesthetic activities such as sports or dance.

In occupational training, kinesthetic learning approaches are used in vocational education to enhance skill development in professions requiring manual dexterity such as surgeons, mechanics and sculptors.

In rehabilitation and therapy, physical activities and body awareness exercises are used in physical therapy and occupational therapy. Therapies like dance therapy or yoga are used for emotional and psychological well-being.

13. Visual-Spatial Intelligence. Visual-Spatial Intelligence is one of the eight intelligences proposed by Howard Gardner in his seminal book *Frames of Mind: The Theory of Multiple Intelligences* [14]. Gardner emphasized that this form of intelligence is distinct and measurable through tasks requiring spatial and visual skills. Shepard and Metzler provide a foundational study in cognitive psychology showing how people manipulate visual-spatial information mentally in their paper *Mental Rotation of Three-Dimensional Objects.* [45]. In the book *Human abilities: their nature and measurement* in a chapter entitled *Spatial Ability and G.* David Lohman [29] explores the relationship between spatial reasoning and general intelligence.

Visual-spatial intelligence is an advanced form of spatial reasoning that involves the capacity to understand and manipulate spatial relationships; It includes the ability to perceive, analyze, and mentally manipulate visual and spatial information. Individuals with strong visual-spatial intelligence excel in tasks that require spatial reasoning, visualization, and recognition of patterns or relationships in space. Specifically, visual-spatial intelligence is the capacity to accurately perceive the visual world, manipulate mental images and spatial constructs, solve problems involving visual-spatial relationships, and create representations of physical spaces, such as maps, designs, or artistic works.

Visual-spatial intelligence is crucial across various professional domains, including engineering, architecture, and science, where it facilitates the conceptualization and development of complex systems. It enhances problem solving and innovation in mathematics and technology and is fundamental in creative fields such as graphic design, animation, sculpture, and photography, contributing to aesthetic understanding and artistic expression. In addition, it plays a critical role in professions requiring spatial navigation, such as piloting, sailing, and urban planning.

For example, *architects* rely on visual-spatial intelligence to design and visualize structures, while *artists* apply it to create detailed and visually compelling works. *Engineers* use it to conceptualize and construct intricate systems, and *pilots* and *surgeons* depend on it for precise spatial awareness. Similarly, *navigators* use it for map reading and directional comprehension. These diverse applications underscore its significance across multiple disciplines.

Visual-spatial intelligence divides into five components as follows: 1) *Mental Visualization* the ability to create and manipulate mental images or scenarios 2) *Artistic and Design Abilities* the talent to draw, paint or create three-dimensional works, 3) *Spatial Reasoning* the ability to understand and navigate spatial relationships, such as distances, orientations, and geometries, 4) *Pattern Recognition*, the ability to identify visual patterns and connections, and 5) *Orientation Skills* a strong sense of direction and the ability to visualize spatial layouts.

In addition to the arts, navigation, and engineering disciplines alluded to above, visual intelligence finds practical use in education. This is specifically referred to by Armstrong in *Multiple Intelligences in the Classroom.* [3] where practical strategies are provided to identify and promote visual-spatial intelligence in educational settings. Newcombe discusses the role of visual-spatial intelligence in STEM fields in *Picture This: Increasing math and science learning by improving spatial thinking* [33]. In schools, the use of visual aids, such as diagrams, maps, and videos, helps improve learning while encouraging activities such as drawing, model building, and puzzles to develop spatial skills.

The following activities are used to improve visual-spatial intelligence: *Practicing Visualization*, participating in exercises like mental rotations or imagining spaces, *Using Spatial Tools*, experiment with maps, blueprints, and 3D models, *Play Games*, participate in puzzles, chess, or video games that require spatial reasoning, *Exploring Art and Design*, practicing drawing, painting, or creating 3D crafts, and *Outdoor Activities*, participating in activities like hiking or orienteering to enhance spatial awareness.

14. Metacognitive intelligence is the ability to understand, monitor, and regulate one's cognitive processes.

Metacognitive Intelligence refers to the ability to think about and regulate one's own cognitive processes. It involves self-awareness of how one learns, plans, and solves problems, as well as the ability to monitor, evaluate, and adapt one's thinking strategies to achieve specific goals.

John Flavell in his 1979 paper*Metacognition and Cognitive Monitoring: A New Area of Cognitive-Developmental Inquiry.* [13] Introduced the concept of metacognition, focusing on its role in learning and cognitive development. Livingston provides a comprehensive overview of metacognition, its components, and practical applications in *Metacognition: An Overview* [28] in which he says that metacognition is often described as "thinking about thinking." Metacognition plays a critical role in learning, decision-making, and problem solving. Although not explicitly included in Howard Gardner's *Theory of Multiple Intelligences*, metacognitive intelligence intersects many cognitive and emotional skills. In 1987, Anne Brown explored metacognition in relation to executive control and learning processes in her paper *Metacognition, Executive Control, Self-Regulation, and Other More Mysterious Mechanisms.* [6].

Metacognitive intelligence can be broken down into two main components:

1. **Metacognitive Knowledge**: Awareness of one's cognitive processes, including knowledge about oneself as a learner, such as recognizing strengths and weaknesses, understanding strategies that can improve learning and problem solving, and awareness of task requirements and how to approach them.
2. **Metacognitive Regulation**: The ability to control and adapt cognitive processes, including setting goals and choosing strategies, tracking progress and identifying challenges, and assessing the effectiveness of strategies and outcomes.

Metacognitive intelligence is characterized by *Self-Awareness*, the ability to understand one's cognitive strengths and limitations, *Strategic Thinking*, knowing how to choose and apply effective learning or problem-solving strategies, *Flexibility* the ability to adapt strategies in response to changing conditions or feedback, and *Goal-Orientation*, the ability to focus on achieving objectives through efficient cognitive regulation.

Metacognition is ubiquitous, and a student who reflects on why they struggled with a math problem and decides to revise using new methods, using a checklist to ensure that all steps in an assignment are completed shows metacognitive intelligence. An engineer who revisits a failed design and brainstorms alternative approaches uses metacognitive intelligence. A manager evaluating team strategies and reallocating resources for better outcomes uses metacognitive intelligence. Tracking expenses and adjusting budgeting methods based on observed spending patterns involve metacognitive intelligence. An athlete analyzing their performance to refine their training regimen is using metacognitive intelligence.

Due to its ubiquity in learning, in work and in play, metacognitive intelligence is important. It is what allows us to make good decisions. High metacognitive intelligence allows for improved learning outcomes as it encourages students to become self-directed learners by identifying and applying effective strategies. It enables better decision making by helping people assess the risks and benefits of actions. It improves problem solving outcomes by encouraging a reflective approach to complex challenges, improving efficiency and effectiveness. It supports resiliency and adaptability by reflecting on previous failures.

Metacognitive intelligence finds applications wherever a reflective mindset is valuable. Paul Pintrich's 2002 article*The Role of Metacognitive Knowledge in Learning, Teaching, and Assessing.* [37] discusses the integration of metacognitive strategies in educational practices.

Teaching students metacognition skills makes them more successful in their studies and beyond. Examples include teaching students to reflect on their learning strategies and adjust based on outcomes, and using metacognitive prompts, such as "What did you learn?" or "How will you approach this differently next time?" In the workplace, training employees to monitor and adapt their problem-solving strategies and encouraging reflective practices for continuous improvement are examples of how ideas from metacognitive intelligence lead to better outcomes for businesses. Healthcare also benefits from efforts to support patients in managing cronic conditions by teaching self-monitoring and planning strategies, and also by helping therapists guide clients in self-reflective and adaptive thinking. Finally, in leadership positions, the use of metacognitive strategies to assess team performance and refine goals and the cultivation of a reflective culture within organizations leads to better outcomes.

As with most dimensions of intelligence, for a dimension that has such impact, there is interest in improving metacognitive intelligence wherever education occurs at all levels, from school to aging. The article by Schraw and Moshman *Metacognitive Theories.* [42] examines how metacognitive theories help individuals understand and regulate their cognitive processes.

15. Adaptive Intelligence. *Adaptive Intelligence* refers to the ability to adjust one's thinking, behavior, and actions to navigate new or changing environments effectively. It involves solving problems and making decisions that align with personal goals, cultural contexts, and social norms, particularly in uncertain or novel situations. Adaptive intelligence is a dynamic process that integrates creativity, critical thinking, emotional intelligence, and resilience.

It is often described as a key element in overall intelligence and complements other forms such as analytical, creative, and emotional intelligence. Adaptive intelligence is linked to evolutionary success, emphasizing the importance of learning and adapting to survive and thrive in complex environments.

This concept emphasizes practical problem solving and thriving in a rapidly changing world, such as technological advances, cultural changes, and environmental challenges.

Adaptive intelligence has its roots in several psychological and cognitive theories. Jean Piaget's book *The origins of intelligence in children* [47] discusses adaptation as a core element of human cognitive development. Piaget highlighted the importance of adaptation through assimilation (integrating new experiences into existing frameworks) and accommodation (adjusting frameworks to incorporate new information). Sternberg discusses the practical and adaptive aspects of intelligence within his triarchic model in his book *Adaptive Intelligence: Sur-*

viving and Thriving in Times of Uncertainty. [50]. John Carroll's book *Human Cognitive Abilities: A Survey of Factor-Analytic Studies.* [7] is a foundational work linking various cognitive abilities, including adaptability, to intelligence. The role of creativity, a key component of adaptive intelligence, in problem solving and innovation is highlighted in Plucker and Makel's *Assessment of Creativity in Education and Psychology* [38].

Adaptive intelligence can be summarized as the capacity to modify strategies or behavior in response to new information or circumstances (flexability), effectively addressing challenges in real-world complex situations (problem solving), balancing individual goals with societal and cultural considerations (social responsibility), and continuously improving and learning from experience (learning orientation).

Adaptive intelligence can be divided into the following components: 1) *Cognitive Flexibility*, the ability to change perspectives and adopt alternative approaches, 2) *Practical Problem-Solving* the focus on actionable solutions to everyday challenges, 3) *Resilience*, the capacity to recover from setbacks and learn from failure, 4) *Cultural Sensitivity* understanding and adapting to diverse cultural norms and expectations, and 4) *Futuristic Thinking* anticipating change and preparing for future uncertainties.

Examples of adaptive intelligence can be found in education, where a student adjusts study methods after realizing that their initial approach was ineffective, or where a student adapts to remote learning environments during a pandemic. In Leadership, where a manager revises strategies in response to sudden market changes or when leading a multicultural team when diverse cultural needs are recognized and addressed. In Everyday Life, learning how to use new technology after a system upgrade or learning to live in a foreign country by adapting to local customs and practices.

Applications of adaptive intelligence can be found in education, where students are taught to approach problems flexibly and think critically, and where learners are prepared for future uncertainties by emphasizing adaptability over memorization. In the workplace, employees can be trained to adapt to technological advancements and changing market dynamics. Leaders may be encouraged to employ leadership styles that are responsive to change and diverse team needs. In healthcare, adaptive intelligence can help patients adjust to chronic illnesses or life-changing diagnoses. Healthcare professionals may be trained to respond effectively to changing medical practices and emergencies.

An individual with high adaptive intelligence is better able to adapt to the changes that life presents.

7 Conclusion

It is clear to the authors that systematic and general intelligence tests that can be applied to agents, human, animal, or computer-based, will be tremendously useful. Our first thought was to harness the over a century of effort in measuring human intelligence, as an efficient starting point. However, three issues argue against this approach. A review of intelligence testing in humans is rather disappointing, showing remarkably modest improvement since the original work by Binet. The second issue is that we need intelligence tests that can be generated instead of published, because prior exposure to intelligence test questions skews the score on a test. Finally, we agree with many psychology critics of human intelligence testing that it ignores many dimensions of intelligence, such as social intelligence.

An alternative place to begin is the many benchmarks for Large Language Models that already exist and continue to be developed. However useful these benchmarks are in comparing and contrasting LLMs, they are of little to no use in measuring intelligence.

More promising are efforts such as ARC-AGI, which take seriously the effort needed to assess the intelligence of AI artifacts.

Any effort to seriously consider general measures of intelligence should deal with two very important issues:

- broadening biological notions of intelligence to include animals like birds, dolphins and great apes
- dealing with the issues that arise from non-biological agents, such as computational artifacts.

This paper has largely focused on a review of where we are in intelligence testing and the many issues and barriers to be considered. In a future paper, we would like to present a draft view of how intelligence testing should be done.

Disclosure of Interests. The authors had no competing interests.

References

1. Albrecht, K.: Social Intelligence: The New Science of Success. Jossey-Bass, San Francisco (2006)
2. Amabile, T.M.: Creativity in Context: Update to the Social Psychology of Creativity. Westview Press, Boulder (1996)
3. Armstrong, T.: Multiple Intelligences in the Classroom, 3rd edn. ASCD, Alexandria (2009)
4. Bandura, A.: Social Learning Theory. Prentice-Hall, Englewood Cliffs (1977)
5. Borba, M.: Building Moral Intelligence: The Seven Essential Virtues that Teach Kids to do the Right Thing. Jossey-Bass, San Francisco (2001)
6. Brown, A.L.: Metacognition, executive control, self-regulation, and other more mysterious mechanisms. Adv. Child Dev. Behav. **21**, 65–116 (1987)
7. Carroll, J.B.: Human Cognitive Abilities: A Survey of Factor-Analytic Studies. Cambridge University Press, Cambridge (1993)
8. Cattell, R.B.: Theory of fluid and crystallized intelligence: a critical experiment. J. Educ. Psychol. **54**(1), 1–22 (1963)
9. Chollet, F.: On the measure of intelligence. CoRR abs/1911.01547 (2019). http://arxiv.org/abs/1911.01547
10. Csikszentmihalyi, M.: Flow: The Psychology of Optimal Experience. Harper & Row, New York (1990)

11. Earley, P.C., Ang, S.: Cultural Intelligence: Individual Interactions Across Cultures. Stanford University Press, Stanford (2003)
12. Emmons, R.A.: The Psychology of Ultimate Concerns: Motivation and Spirituality in Personality. Guilford Press, New York (1999)
13. Flavell, J.H.: Metacognition and cognitive monitoring: a new area of cognitive-developmental inquiry. Am. Psychol. **34**(10), 906–911 (1979)
14. Gardner, H.: Frames of Mind: The Theory of Multiple Intelligences. Basic Books, New York (1983)
15. Gardner, H.: Intelligence Reframed: Multiple Intelligences for the 21st Century. Basic Books, New York (1999)
16. Georgeon, O.L., Marshall, J.: Demonstrating sensemaking emergence in artificial agents: a method and an example. Int. J. Mach. Conscious. **5**, 131–144 (2013). https://hal.science/hal-01339177
17. Gilligan, C.: In a Different Voice: Psychological Theory and Women's Development. Harvard University Press, Cambridge (1982)
18. Goleman, D.: Emotional Intelligence: Why It Can Matter More Than IQ. Bantam Books, New York (1995)
19. Goleman, D.: Social Intelligence: The New Science of Human Relationships. Bantam Books, New York (2006)
20. Gottfredson, L.S.: Dissecting practical intelligence theory: its claims and evidence. Intelligence **31**(4), 343–397 (2003)
21. Horn, J.L.: The theory of fluid and crystallized intelligence in relation to concepts of cognitive psychology and aging in adulthood. Aging Hum. Dev. **15**(2), 109–121 (1982)
22. Horn, J.L., Cattell, R.B.: Refinement and test of the theory of fluid and crystallized intelligence. J. Educ. Psychol. **57**(5), 253–270 (1966)
23. Jones, C.R., Bergen, B.K.: People cannot distinguish GPT-4 from a human in a turing test (2024). https://arxiv.org/abs/2405.08007
24. Kambhampati, S., et al.: LLMs can't plan, but can help planning in LLM-modulo frameworks (2024). https://arxiv.org/abs/2402.01817
25. Kohlberg, L.: Essays on Moral Development, Volume 1: The Philosophy of Moral Development. Harper & Row, San Francisco (1984)
26. Lennick, D., Kiel, F.: Moral Intelligence: Enhancing Business Performance and Leadership Success. Wharton School Publishing, Upper Saddle River (2005)
27. Livermore, D.: Leading with Cultural Intelligence: The New Secret to Success. AMACOM, New York (2010)
28. Livingston, J.A.: Metacognition: an overview (1997). https://fkeste01.wordpress.com/j-livingston/
29. Lohman, D.F.: Spatial ability and G. In: Frearson, J., Smith, R.M. (eds.) Human Abilities: Their Nature and Measurement, pp. 97–116. Lawrence Erlbaum Associates, Hillsdale (1993)
30. Matthews, G., Zeidner, M., Roberts, R.D.: Emotional Intelligence: Science and Myth. MIT Press, Cambridge (2002)
31. McArdle, J.J., Woodcock, R.W.: Human Abilities and Intelligence: Structure, Measurement, and Development. Psychological Press, Mahwah (1998)
32. Neisser, U., et al.: Intelligence: knowns and unknowns. Am. Psychol. **51**(2), 77–101 (1996)
33. Newcombe, N.S.: Picture This: Increasing Math and Science Learning by Improving Spatial Thinking. American Psychological Association, Washington, DC (2010)

34. Ng, K.Y., Van Dyne, L., Ang, S.: From experience to experiential learning: cultural intelligence as a learning capability for global leader development. Acad. Manag. Learn. Educ. **8**(4), 511–526 (2009)
35. Petrides, K.V., Furnham, A.: Trait emotional intelligence: psychometric investigation with reference to established trait taxonomies. Eur. J. Pers. **15**(6), 425–448 (2001)
36. Phan, L., et al.: Humanity's last exam. arXiv (2025)
37. Pintrich, P.R.: The role of metacognitive knowledge in learning, teaching, and assessing. Theory Into Pract. **41**(4), 219–225 (2002)
38. Plucker, J.A., Makel, M.C.: Assessment of creativity in education and psychology. In: Kaufman, J.C., Sternberg, R.J. (eds.) The Cambridge Handbook of Creativity, pp. 48–73. Cambridge University Press, Cambridge (2010)
39. Runco, M.A., Pritzker, S.R. (eds.): Encyclopedia of Creativity. Academic Press, San Diego (1999)
40. Salovey, P., Mayer, J.D.: Emotional intelligence. Imag. Cogn. Pers. **9**(3), 185–211 (1990)
41. Salthouse, T.A.: What and when of cognitive aging. Curr. Dir. Psychol. Sci. **13**(4), 140–144 (2004)
42. Schraw, G., Moshman, D.: Metacognitive theories. Educ. Psychol. Rev. **7**(4), 351–371 (1995)
43. Searle, J.: Minds, brains, and programs. Behav. Brain Sci. **3**, 417–457 (1980)
44. Shapiro, L.: Embodied Cognition. Routledge, New York (2014)
45. Shepard, R.N., Metzler, J.: Mental rotation of three-dimensional objects. Science **171**(3972), 701–703 (1971)
46. Spearman, C.: General intelligence, objectively determined and measured. Am. J. Psychol. **15**(2), 201–292 (1904)
47. Sternberg, R.J.: Beyond IQ: A Triarchic Theory of Human Intelligence. Cambridge University Press, Cambridge (1985)
48. Sternberg, R.J.: Intelligence, Information Processing, and Analogical Reasoning: The Componential Analysis of Human Abilities. Psychology Press, New York (1997)
49. Sternberg, R.J.: Practical Intelligence in Everyday Life. Cambridge University Press, Cambridge (2000)
50. Sternberg, R.J.: Adaptive Intelligence: Surviving and Thriving in Times of Uncertainty. Cambridge University Press, Cambridge (2021). https://doi.org/10.1017/9781316650554
51. Sternberg, R.J., Kaufman, J.C.: The Evolution of Intelligence: Are Humans the Only Creatures Who Ask 'Why?'. Lawrence Erlbaum Associates, Mahwah (2004)
52. Sternberg, R.J., Wagner, R.K., Okagaki, L.: Practical intelligence: the nature and role of tacit knowledge in work and at school. In: Reese, H.W., Puckett, J.M. (eds.) Advances in Lifespan Development. Academic Press (1993)
53. Thomas, D.C., Inkson, K.: Cultural Intelligence: Surviving and Thriving in the Global Village, 3rd edn. Berrett-Koehler Publishers, Oakland (2017)
54. Thorndike, E.L.: Intelligence and its uses. Harper's Mag. **140**(801), 227–235 (1920)
55. Turing, A.M.: Computing machinery and intelligence. Mind **59**(236), 433–460 (1950). http://www.jstor.org/stable/2251299
56. Van Dyne, L., Ang, S., Koh, C.: Development and validation of the CQS (cultural intelligence scale). In: Handbook of Cultural Intelligence: Theory, Measurement, and Applications, pp. 16–38. M.E. Sharpe (2009)

57. Warwick, K., Shah, H.: Can machines think? A report on turing test experiments at the royal society. J. Exp. Theor. Artif. Intell. **28**(6), 989–1007 (2016). https://doi.org/10.1080/0952813X.2015.1055826
58. Waterhouse, L.: Multiple intelligences, the mozart effect, and emotional intelligence: a critical review. Educ. Psychol. **41**(4), 207–225 (2006)
59. Wink, M.: Intelligence in the Animal Kingdom, pp. 87–103. Springer, Cham (2022). https://doi.org/10.1007/978-3-031-04198-3_4

Author Index

G
Georgeon, Olivier L. 1, 62, 88

K
Kommrusch, Steve 53

L
Laddaga, Robert 121

M
Minsky, Henry 53

P
Perotto, Filipo S. 62

R
Robertson, Paul 1, 62, 88, 121

S
Sheikhlar, Arash 62

T
Thórisson, Kristinn R. 10, 62

W
Wang, Pei 10

P. Robertson and O. Georgeon (Eds.): IWSSL 2024, LNCS 15457, p. 151, 2025.
https://doi.org/10.1007/978-3-031-96325-4

The manufacturer's authorised representative in the EU is Springer Nature Customer Service Centre GmbH, Europaplatz 3, 69115 Heidelberg, Germany. If you have any concerns regarding our products, please contact ProductSafety@springernature.com

Printed and bound by CPI Group (UK) Ltd, Croydon, CR0 4YY
01/10/2026
02225984-0005